Evaluation of Demonstration Test Results of Alternative Technologies for Demilitarization of Assembled Chemical Weapons

A Supplemental Review

Committee on Review and Evaluation of Alternative Technologies
for Demilitarization of Assembled Chemical Weapons

Board on Army Science and Technology

Commission on Engineering and Technical Systems
National Research Council

NATIONAL ACADEMY PRESS
Washington, DC

NOTICE: The project that is the subject of this report was approved by the Governing Board of the National Research Council, whose members are drawn from the councils of the National Academy of Sciences, the National Academy of Engineering, and the Institute of Medicine. The members of the committee responsible for the report were chosen for their special competencies and with regard for appropriate balance.

This is a report of work supported by Contract **DAAM01-97-C-0015** between the U.S. Army and the National Academy of Sciences. Any opinions, findings, conclusions, or recommendations expressed in this publication are those of the author(s) and do not necessarily reflect the view of the organizations or agencies that provided support for the project.

International Standard Book Number 0-309-06897-5

Limited copies are available from:

Board on Army Science and Technology
National Research Council
2101 Constitution Avenue, N.W.
Washington, DC 20418
(202) 334-3118

Additional copies of this report are available from:

National Academy Press
2101 Constitution Avenue, N.W., Lockbox 285
Washington, DC 20055
(800) 624-6242 or (202) 334-3313 (in the
Washington metropolitan area)
http://www.nap.edu

Printed in the United States of America.

THE NATIONAL ACADEMIES

National Academy of Sciences
National Academy of Engineering
Institute of Medicine
National Research Council

The **National Academy of Sciences** is a private, nonprofit, self-perpetuating society of distinguished scholars engaged in scientific and engineering research, dedicated to the furtherance of science and technology and to their use for the general welfare. Upon the authority of the charter granted to it by the Congress in 1863, the Academy has a mandate that requires it to advise the federal government on scientific and technical matters. Dr. Bruce M. Alberts is president of the National Academy of Sciences.

The **National Academy of Engineering** was established in 1964, under the charter of the National Academy of Sciences, as a parallel organization of outstanding engineers. It is autonomous in its administration and in the selection of its members, sharing with the National Academy of Sciences the responsibility for advising the federal government. The National Academy of Engineering also sponsors engineering programs aimed at meeting national needs, encourages education and research, and recognizes the superior achievements of engineers. Dr. William A. Wulf is president of the National Academy of Engineering.

The **Institute of Medicine** was established in 1970 by the National Academy of Sciences to secure the services of eminent members of appropriate professions in the examination of policy matters pertaining to the health of the public. The Institute acts under the responsibility given to the National Academy of Sciences by its congressional charter to be an adviser to the federal government and, upon its own initiative, to identify issues of medical care, research, and education. Dr. Kenneth I. Shine is president of the Institute of Medicine.

The **National Research Council** was organized by the National Academy of Sciences in 1916 to associate the broad community of science and technology with the Academy's purposes of furthering knowledge and advising the federal government. Functioning in accordance with general policies determined by the Academy, the Council has become the principal operating agency of both the National Academy of Sciences and the National Academy of Engineering in providing services to the government, the public, and the scientific and engineering communities. The Council is administered jointly by both Academies and the Institute of Medicine. Dr. Bruce M. Alberts and Dr. William A. Wulf are chairman and vice chairman, respectively, of the National Research Council.

COMMITTEE ON REVIEW AND EVALUATION OF ALTERNATIVE TECHNOLOGIES FOR DEMILITARIZATION OF ASSEMBLED CHEMICAL WEAPONS

ROBERT A. BEAUDET, *chair*, University of Southern California, Los Angeles
RICHARD J. AYEN, Waste Management, Inc. (retired), Jamestown, Rhode Island
JOAN B. BERKOWITZ, Farkas Berkowitz and Company, Washington, D.C.
NOSA O. EGIEBOR, Tuskegee University, Tuskegee, Alabama
WILLARD C. GEKLER, EQE International/PLG, Irvine, California
HANK C. JENKINS-SMITH, University of New Mexico, Albuquerque
JOHN L. MARGRAVE, Rice University, Houston, Texas
WALTER G. MAY, University of Illinois (retired), Urbana
KIRK E. NEWMAN, Naval Surface Warfare Center, Indian Head Division, Yorktown, Virginia
JIMMIE C. OXLEY, University of Rhode Island, Kingston
WILLIAM R. RHYNE, H&R Technical Associates, Inc., Oak Ridge, Tennessee
STANLEY I. SANDLER, University of Delaware, Newark
WILLIAM R. SEEKER, General Electric Energy and Environmental Research Corporation, Irvine, California
LEO WEITZMAN, LVW Associates, Inc., West Lafayette, Indiana

Board on Army Science and Technology Liaison

WILLIAM H. FORSTER, *chair*, Northrop Grumman Corporation, Baltimore, Maryland

Staff

BRUCE A. BRAUN, Study Director
HARRISON T. PANNELLA, Research Associate
JACQUELINE CAMPBELL-JOHNSON, Senior Project Assistant

BOARD ON ARMY SCIENCE AND TECHNOLOGY

WILLIAM H. FORSTER, *chair*, Northrop Grumman Corporation, Baltimore, Maryland
THOMAS L. MCNAUGHER, *vice chair*, RAND Corporation, Washington, D.C.
ELIOT A. COHEN, School of Advanced International Studies, Johns Hopkins University, Washington, D.C.
RICHARD A. CONWAY, Union Carbide Corporation (retired), Charleston, West Virginia
GILBERT F. DECKER, Walt Disney Imagineering, Glendale, California
PATRICK F. FLYNN, Cummins Engine Company, Inc., Columbus, Indiana
EDWARD J. HAUG, NADS and Simulation Center, University of Iowa, Iowa City
ROBERT J. HEASTON, Guidance and Control Information Analysis Center (retired), Naperville, Illinois
ELVIN R. HEIBERG, Heiberg and Associates, Inc., Mason Neck, Virginia
GERALD J. IAFRATE, University of Notre Dame, Notre Dame, Indiana
DONALD R. KEITH, Cypress International, Alexandria, Virginia
KATHRYN V. LOGAN, Georgia Institute of Technology, Atlanta
JOHN E. MILLER, Oracle Corporation, Reston, Virginia
JOHN H. MOXLEY, Korn/Ferry International, Los Angeles, California
STEWART D. PERSONICK, Drexel University, Philadelphia, Pennsylvania
MILLARD F. ROSE, NASA Marshall Space Flight Center, Huntsville, Alabama
GEORGE T. SINGLEY, III, Hicks and Associates, Inc., McLean, Virginia
CLARENCE G. THORNTON, Army Research Laboratories (retired), Colts Neck, New Jersey
JOHN D. VENABLES, Venables and Associates, Towson, Maryland
JOSEPH J. VERVIER, ENSCO, Inc., Melbourne, Florida
ALLEN C. WARD, Ward Synthesis, Inc., Ann Arbor, Michigan

Staff

BRUCE A. BRAUN. Director
MICHAEL A. CLARKE, Associate Director
MARGO L. FRANCESCO, Staff Associate
CHRIS JONES, Financial Associate
DEANNA SPARGER, Senior Project Assistant

Preface

The United States has been in the process of destroying its chemical munitions for over a decade. The U.S. Army, with expertise from numerous bodies including the National Research Council (NRC), originally decided to use incineration as the method of destruction at all storage sites. However, citizens in states with storage sites have opposed incineration on the grounds that it is impossible to determine the exact nature of the effluents, in particular, effluents from the stacks. Nevertheless, the Army has continued to pursue incineration at most sites. In the last few years, influenced by growing public opposition to incineration and after numerous studies, including a 1996 study by the NRC entitled *Review and Evaluation of Alternative Chemical Disposal Technologies*, the Army is developing a chemical neutralization process to destroy chemical agents stored only in bulk ton containers at two sites: VX at Newport, Indiana, and mustard (HD) at Aberdeen Maryland.

Pursuaded by public opposition to incineration at the Lexington, Kentucky, and Pueblo, Colorado, sites, Congress in 1996 enacted Public Law 104-201 instructing the Department of Defense (DOD) to "conduct an assessment of the chemical demilitarization program for destruction of assembled chemical munitions and of the alternative demilitarization technologies and processes (other than incineration) that could be used for the destruction of the lethal chemical agents that are associated with these munitions." The Army established a Program Manager for Assembled Chemical Munitions Assessment (PMACWA) to respond to this instruction. Unlike prior activities, the PMACWA involved the public in every aspect of the program including the procurement process. A nonprofit organization, the Keystone Center, was hired to facilitate public involvement.

After requesting and receiving proposals from industry for complete technology packages to destroy stored assembled chemical weapons, the Army initially selected seven industry teams, denoted as technology providers in this report. In later selections, these seven were reduced to six, and then three to proceed to the demonstration phase of the assessment program. When the NRC's Committee on Review and Evaluation of Alternative Technologies for Demilitarization of Assembled Chemical Weapons (ACW Committee) first report was written, the committee did not have the benefit of evaluating the results of the demonstrations.

Subsequently, the PMACWA requested that the committee evaluate both the technology providers' test reports and the Army's evaluations to determine if the demonstrations changed the committee's earlier findings or recommendations. This report is a supplemental review evaluating the impact of the three demonstration tests on the committee's original findings and recommendations.

I wish to acknowledge with great gratitude the members of the ACW Committee who have continued to serve as volunteers throughout this extended study and who completed this supplemental study in the relatively short time allocated by the PMACWA. They provided the necessary expertise in chemical processing, permitting and regulations, energetic materials and public acceptance to continue this task. I remain, by far, the least capable of this group.

The committee recognizes and appreciates the assistance of the Army ACWA team, which provided support and the necessary reports. We also appreciate the openness and the cordiality of the technology providers.

A study such as this requires extensive support. We are all indebted to the NRC staff for their logistic support. I would particularly like to acknowledge the close working relationship between the committee and Bruce Braun, who undertook the task of acting study director along with his other duties as director of the NRC Board on Army Science and Technology. Mr. Braun also provided the resources and staff to complete this study in record time for an NRC report. The efforts of Harrison Pannella, who acted as assistant study director, were invaluable. He put in long hours on evenings and weekends to prepare, edit, and format this report. In addition, Rebecca Lucchese and Jacqueline Johnson provided logistic support to the committee, allowing us to concentrate on our task. Also, an acknowledgement is due for Carol Arenberg, who edited the final draft of the report. Everyone worked under a short deadline and great stress during a period that included a holiday season.

I gratefully acknowledge the support of my colleagues in the Chemistry Department at the University of Southern California, who willingly assumed my teaching duties while I traveled on behalf of this study.

Robert A. Beaudet, *chair*
Committee on Review and Evaluation of
Alternative Technologies for Demilitarization
of Assembled Chemical Weapons

Acknowledgment

This report has been reviewed by individuals chosen for their diverse perspectives and technical expertise, in accordance with procedures approved by the National Research Council's Report Review Committee. The purpose of this independent review is to provide candid and critical comments that will assist the authors and the NRC in making the published report as sound as possible and to ensure that the report meets institutional standards for objectivity, evidence, and responsiveness to the study charge. The contents of the review comments and draft manuscript remain confidential to protect the integrity of the deliberative process. We wish to thank the following individuals for their participation in the review of this report:

Richard Magee, New Jersey Institute of Technology
Raymond McGuire, Lawrence Livermore National Laboratory
Royce Murray, University of North Carolina
Robert Olson, consultant
George Parshall, E.I. DuPont de Nemours & Company
Janice Phillips, Lehigh University
Martin Sherwin, ChemVen Group, Inc.

While the individuals listed above have provided many constructive comments and suggestions, responsibility for the final content of this report rests solely with the authoring committee and the NRC.

Contents

EXECUTIVE SUMMARY 1

1 INTRODUCTION 4
Background, 4
Role of the National Research Council, 5
Statement of Task, 5
Scope of This Study, 5
Organization of This Report, 5

2 BURNS AND ROE PLASMA ARC PROCESS 6
Plasma Waste Converter, 6
Energetics Campaign, 6
Dunnage and Secondary Waste Campaign, 7
Agent Campaign, 7
Projectile Heel Campaign, 8
Review of Previous Committee Findings, 8
Safety Issues, 10
Reevaluation of Steps Required for Implementation, 11
Supplemental Findings, 11

3 GENERAL ATOMICS TECHNOLOGY PACKAGE 12
Energetics Rotary Hydrolyzer, 12
Dunnage Shredding/Hydropulping System, 13
Supercritical Water Oxidation System, 14
Safety Concerns, 15
Effluent Characterization, 16
Reevaluation of Steps Required for Implementation, 16
Supplemental Findings and Recommendations, 17

4 PARSONS-ALLIEDSIGNAL WHEAT PROCESS 18
Munitions Cutting and Fluid Mining, 18
Biotreatment Systems, 19
Biotreatment System for Mustard Hydrolysate, 19
Biotreatment System for Nerve Agent Hydrolysates, 20
Catalytic Oxidation, 22
Catalytic Oxidation Unit for Mustard, 22
Catalytic Oxidation Unit for Nerve Agent, 22

Metal Parts Treater, 22
Safety Concerns, 23
Reevaluation of Steps Required for Implementation, 23
Review of Previous Committee Findings, 24
Supplemental Findings and Recommendation, 24

5 UPDATE OF GENERAL FINDINGS AND RECOMMENDATIONS 26
Review of Earlier Findings and Recommendations, 26
Supplemental General Findings, 28

REFERENCES 30

APPENDIXES

A FINDINGS AND RECOMMENDATIONS FROM THE 1998 REPORT ON SUPERCRITICAL WATER OXIDATION 31

B BIOGRAPHICAL SKETCHES OF COMMITTEE MEMBERS 34

List of Figures and Tables

FIGURES

4-1 Demonstration test unit for treatment of HD/tetrytol hydrolysate, 19
4-2 Demonstration test unit for treatment of GB/Comp B hydrolysate, 21

TABLE

ES-1 Summary Evaluation of the Maturity of Demonstrated Unit Operations and Processes, 2
5-1 Summary Evaluation of the Maturity of Demonstrated Unit Operations and Processes, 29

Acronyms

ACWA	Assembled Chemical Weapons Assessment (program)
ARAR	appropriate, relevant, and applicable rule
BOD	biological oxygen demand
CAA	Clean Air Act
CAMDS	Chemical Agent Munitions Disposal System
CATOX	catalytic oxidation
CFM	cubic feet per minute
COD	chemical oxygen demand
CSTR	continuously stirred tank reactor
DAAMS	depot area air monitoring system
DMMP	dimethyl methyl phosphonate
DOD	U.S. Department of Defense
DPE	demilitarization protective ensemble (suit)
DRE	destruction and removal efficiency
DSHS	dunnage shredding/hydropulping system
EDC	energetics deactivation chamber
EMPA	ethyl methylphosphonic acid
EPA	Environmental Protection Agency
ERH	energetics rotary hydrolyzer
GB	type of nerve agent
GC	gas chromatography
GC/MS	gas chromatography/mass spectrometry
HD	distilled mustard agent
HEPA	high-efficiency particulate air
HRA	health risk assessment
ICB	immobilized cell biotreatment
IMPA	isopropyl methylphosphonic acid

M	molar concentration
MPT	metal parts treater
NRC	National Research Council
PCG	plasma converted gas
PMACWA	Program Manager for Assembled Chemical Weapons Assessment
ppm_v	parts per million (volumetric)
PWC	plasma waste converter
RCRA	Resource Conservation and Recovery Act
RDX	cyclotrimethylenetrinitramine
RFP	request for proposal
scf	standard cubic feet
SCWO	supercritical water oxidation
TCLP	toxicity characteristic leachate procedure
TNT	trinitrotoluene
TWA	time weighted average
UV	ultraviolet
VOC	volatile organic compound
VX	type of nerve agent
WHEAT	water hydrolysis of explosives and agent technology
3X	level of decontamination (suitable for transport for further processing)
5X	level of decontamination (suitable for commercial release)

Executive Summary

In 1996, the U.S. Congress enacted two laws, Public Law 104-201 (authorization legislation) and Public Law 104-208 (appropriation legislation), mandating that the U.S. Department of Defense (DOD) conduct an assessment of alternative technologies to the baseline incineration process for the demilitarization of assembled chemical munitions. In December 1996, DOD appointed Mr. Michael Parker, Technical Director of the Soldier Biological Chemical Command, to be the program manager for assembled chemical weapons assessment (PMACWA). The program manager published a request for proposals for the complete destruction of assembled chemical weapons. On July 29, 1998, three technology packages were selected for the demonstration phase of the ACWA program. Constrained by both time and resources, the PMACWA selected the unit operations deemed "most critical [and] least proven" for demonstration testing.

The PMACWA had previously requested that the National Research Council (NRC) perform and publish an independent evaluation of the seven technologies packages that had been selected during earlier phases of the Assembled Chemical Weapons Assessment (ACWA) program and deliver a report by September 1, 1999. However, to meet that deadline, the NRC Committee on Review and Evaluation of Alternative Technologies for Demilitarization of Assembled Chemical Weapons (ACW Committee) had to terminate its data-gathering activities on March 15, 1999, prior to the completion of demonstration tests. In September 1999, the PMACWA requested that the ACW Committee examine the reports of the demonstration tests and determine if the results changed the committee's original findings, recommendations, and comments. This report documents the committee's reassessment of the findings and recommendations in the original report, *Review and Evaluation of Alternative Technologies for Demilitarization of Assembled Chemical Weapons.*

In this supplemental report, the committee limited its review to the demonstration test reports prepared by the technology providers and the PMACWA's *Supplemental Report to Congress,* which included the PMACWA's technical evaluation of the tests as a separate appendix. The committee limited its evaluation to the effects of the demonstration test results on the earlier report.

The three technology demonstrations are reviewed in separate chapters in this report; in each chapter, the demonstrated unit operations are considered one at a time. Following a short description of the demonstration tests and commentary by the committee, the findings and recommendations from the original report that bear on the demonstrations are then evaluated. In general, very few of the original findings and recommendations were influenced by the demonstrations. In some cases, the original findings and recommendations were confirmed. A number of new findings and recommendations resulted from the demonstrations, however, and these are presented below.

SUPPLEMENTAL FINDINGS AND RECOMMENDATIONS

Burns and Roe Demonstration Tests

Finding BR-1. The plasma torch apparatus, as demonstrated by the Burns and Roe team, is not qualified for further consideration for the demilitarization of assembled chemical weapons. The torch design appears to be unreliable for extended use. Furthermore, the design increases the possibility of a catastrophic water leak, which could produce a significant increase in pressure in the plasma waste converter (PWC), and possibly cause an explosion, which, in turn, could expose personnel to chemical agent. Moreover, the effectiveness of the monitoring and control sensors was not demonstrated.

Finding BR-2. Even after more than a year of research and development, the technology provider has not been able to

TABLE ES-1 Summary Evaluation of the Maturity of Demonstrated Unit Operations and Processes[a]

	Hydrolysates			Agent Munitions			
Unit Operation/Process	VX/GB	HD	Energetics	VX/GB	HD	Energetics	Other
Burns and Roe							
Plasma waste converter[b]	C	C	D	D	D	E	C[c, d,e]
General Atomics							
Hydrolysis				A	A		
Rotary hydrolyzer						C	
Shredding/hydropulping							A[c]
SCWO	B	B	C				C[c]
Parsons-AlliedSignal							
Munitions accessing				B	B	B	
Hydrolysis				A	A	C	
Biotreatment	D	A	A				
Catalytic oxidation							B[e]
Metal parts treater				B	B	D	B[d]

Note: Environmental and safety issues were considered in assigning maturity categorizations. Schedule and cost issues were not considered.

[a] The letter designations are defined as follows (a blank space indicates categorization was not applicable for that material).

A Demonstration provides sufficient information to allow moving forward to full-scale design with reasonable probability of success.

B Demonstration provides sufficient information to allow moving forward to the pilot stage with reasonable probability of success.

C Demonstration indicates that unit operation or process requires additional refinement and additional demonstration before moving forward to pilot stage.

D Not demonstrated; more R&D required.

E Demonstrated unit operation or process is inappropriate for treatment.

[b]Includes integrated gas polishing system to support demonstration

[c]Dunnage

[d]Metal parts

[e]Effluents

show that its small PWC can adequately destroy agent simulants or that nitrogen is the best gas to use for the plasma feed. If oxygen leaks into the reactor, it could react violently with hydrogen. If air were used for the plasma feed gas, regulatory compliance issues would arise, as well as questions of public acceptance.

Finding BR-3. In the absence of any data for processing effluents from agent runs, the committee could not validate the ability of the proposed system to handle and stabilize effluent products arising from agent processing.

General Atomics Demonstration Tests

Finding GA-1. Testing on the hydrolysis of energetic materials contaminated with agent will be necessary before a full-scale system is built and operated.

Finding GA-2. Testing will be required to verify that the larger diameter supercritical water oxidation (SCWO) reactor feed nozzles will be capable of accepting the dunnage material as shredded (i.e., without additional classification and segregation) and that the reactor will perform reliably under these conditions.

Recommendation GA-1. Operation of the size reduction and slurrying system, and long-term operation of the supercritical water oxidation (SCWO) reactor with slurry, should be conducted before proceeding with a full-scale system.

Recommendation GA-2. Before construction of a full-scale supercritical water oxidation (SCWO) system, additional evaluations of construction materials and fabrication techniques will be necessary because corrosion and plugging prevent continuous operation with the present design. If the new construction materials do not solve these problems, then alternative SCWO reactor designs should be investigated.

Recommendation GA-3. To determine the operability of the supercritical water oxidation (SCWO) reactor and the reliability of the materials of construction, long duration runs of a SCWO reactor should be conducted with slurry, with energetics hydrolysate, and with agent hydrolysate before full-scale implementation proceeds.

Recommendation GA-4. The efficacy and safety of the additional step to remove aluminum hydroxide from the hydrolysate produced from rocket propellants should be evaluated prior to construction of a full-scale supercritical water oxidation (SCWO) system.

Recommendation GA-5. Decontamination of solid munitions materials by flushing and immersion should be demonstrated prior to full-scale implementation.

Recommendation GA-6. The air emissions data from the demonstration tests should be used in a screening risk assessment. The results of the air effluent samples should be subject to (1) a human health risk assessment following the Human Health Risk Assessment Protocol (HHRAP) for Hazardous Waste Combustion Facilities from the Environmental Protection Agency (EPA) [EPA530-D-98-001(A,B,C)], and (2) an ecological risk assessment following a protocol that will be released by EPA in the very near future.

Parsons-AlliedSignal Demonstration Tests

Finding PA-1. The mustard demonstration tests were very encouraging and showed that the process is ready for the next scale-up.

Finding PA-2. The nerve agent demonstration tests had serious problems. However, if the previous tests at the technology provider's laboratory and the results of the demonstration tests are combined, the aggregate results are inconclusive. The reason for the poor demonstration results might be as simple as poor aeration in the bioreactor (see Recommendation PA-1).

Recommendation PA-1. Before proceeding to a further scale-up of GB and VX biotreatment processing, the committee recommends that the following steps be taken:

- The biotreatment process should be examined carefully at bench scale to determine the factors that are critical to success.
- An investigation of analytical techniques should be undertaken to provide more reliable process information.

Supplemental General Findings

The results of the demonstration tests did not significantly affect the committee's original general findings and recommendations and, in some cases, confirmed them. The committee's review of the results of the demonstration tests, however, led to the following new general findings.

General Finding 1. Based on the committee's assessment of the maturity of the various unit operations (as summarized in Table ES-1), none of the three technology packages is ready for *integrated* pilot programming, although certain unit operations are sufficiently mature to bypass pilot testing (e.g., hydrolysis of agent).

General Finding 2. The demonstration tests were not operated long enough to demonstrate reliability and long-term operation.

General Finding 3. The committee reiterates that none of the unit operations has been integrated into a complete system. The lack of integration remains a major concern as a significant obstacle to full-scale implementation.

1

Introduction

BACKGROUND

In 1996, the U.S. Congress enacted two laws, Public Law 104-201 (authorization) and Public Law 104-208 (appropriation), mandating that the U.S. Department of Defense (DOD) conduct an assessment of alternative technologies to the baseline incineration process for the demilitarization of assembled chemical weapons and that not less than two technologies be demonstrated. The law included the following stipulations:

- All funds for the construction of destruction facilities at Blue Grass Depot in Richmond, Kentucky, and at Pueblo Chemical Depot in Pueblo, Colorado, should be frozen.
- DOD should select a program manager who was not and had never been associated with the ongoing incineration destruction.
- DOD should "coordinate" with the National Research Council.

In December 1996, DOD appointed Michael Parker, technical director of the Soldier Biological Chemical Command, to be the program manager for the Assembled Chemical Weapons Assessment (ACWA) Program (PMACWA). On July 28, 1997, after organizing a staff and establishing a program plan, the PMACWA published a Request for Proposals (RFP) for a "total system solution" for the destruction of assembled chemical weapons without using incineration. Twelve proposals were submitted in September 1997. Of these, seven were found to have proposed total system solutions and to have passed the threshold requirements stipulated in the RFP. On July 29, 1998, after an elaborate multitiered selection process, three technology packages were selected for demonstration testing. Detailed descriptions of the selection process and all seven technologies are available in the PMACWA's two annual reports to Congress (DOD, 1997, 1998).

Constrained by both time and budgetary resources, the PMACWA identified unit operations for the three technology packages that were "most critical [and] least proven" for the demonstration tests. These unit operations had not been previously used in the disposal of chemical munitions, nor had they been integrated into a complete system for this application. Two of the three technology packages use base hydrolysis as the primary treatment step to destroy agent and energetic materials. Because most of the uncertainties concerning these technology packages pertain to the secondary treatment of products from the primary treatment step, the PMACWA provided hydrolysates for nerve agents GB and VX and mustard agent HD for testing. Approximately 1,100 gallons of GB hydrolysate and 400 gallons of VX hydrolysate were produced at the Army's Chemical Agent Munitions Disposal System (CAMDS) experimental facility at the Deseret Chemical Depot in Utah. Approximately 4,200 gallons of HD hydrolysate were produced at the Army's Aberdeen Proving Ground in Maryland. The agent hydrolysates provided a representative feedstock for the demonstration tests and enabled characterization of the intermediate product stream for residual agent, including Schedule 2 compounds (agent precursor compounds as defined by the international Chemical Weapons Convention).

Various types and amounts of energetic materials contained in the weapons were reacted with caustic solutions similar to those specified in the technology package proposals of the respective providers. These materials were made available for the demonstrations. Unit operations of the three technology packages were set up, and systemization (preoperational testing) was conducted from January to March 1999. The actual demonstrations began in March 1999 and were completed in May 1999. The technology providers submitted their reports on the demonstration tests to the PMACWA on June 30, 1999 (Burns and Roe, 1999a; General Atomics, 1999a; Parsons-AlliedSignal, 1999a). The PMACWA used these reports and other information to prepare a *Supplemental Report to Congress*, which was submitted on September 30, 1999 (DOD, 1999a).

The committee commends the PMACWA and his staff, as well as the support contractors and technology providers, for completing the demonstrations within the very tight time schedule. The committee recognizes that everyone involved worked long hours, including weekends, to fulfill their tasks.

ROLE OF THE NATIONAL RESEARCH COUNCIL

The PMACWA requested that the National Research Council (NRC) perform and publish an independent evaluation of the technologies by September 1, 1999, a month before the Army's report to Congress was due. The NRC and DOD reached agreement on the Statement of Task in March 1997, and the study was officially begun on May 27, 1997. The committee chose to evaluate all seven technology packages that had passed the threshold requirements stipulated in the RFP. The Statement of Task did not require that the NRC recommend a best technology or compare any of the technologies to the baseline incineration process in use at some storage sites. Although members of the committee visited the demonstration sites prior to systemization of the unit operations in January 1999, in order to produce its final report by September 1, 1999, data-gathering activities had to be terminated on March 15, 1999, prior to receiving the results of the demonstration tests. The committee's report was submitted for peer review on May 1, 1999, and was released to the sponsor and the public on August 25, 1999 (NRC, 1999).

In September 1999, the PMACWA requested that the tenure of the committee be extended to review the results of the demonstrations. The committee was asked to determine if and how the demonstration results affected the committee's commentary, findings, and recommendations, as well as the steps required for implementation (NRC, 1999). In October 1999, the committee began its evaluation of the results of the demonstrations and a determination of the impact of these results on its initial report. The present report is an addendum to the initial report documenting the committee's review of the demonstration test results and the impact of those results on its initial report.

STATEMENT OF TASK

The Statement of Task for this report is as follows:

> At the request of the DOD's Program Manager for Assembled Chemical Weapons Assessment (PMACWA), the NRC Committee on Review and Evaluation of Alternative Technologies for Demilitarization of Assembled Chemical Weapons will continue its independent scientific and technical assessment of the three demonstrated alternative technologies for assembled chemical weapons located at the U.S. chemical weapons storage sites. The continuation of the NRC study will involve the review and evaluation of the demonstration results from the Burns and Roe, General Atomics, and Parsons-AlliedSignal tests performed by the PMACWA. The specific tasks to be performed are:
>
> - use the following as the basis of information:
>   - — PMACWA's *Supplemental Report to Congress* issued September 30, 1999, and the "Technical Evaluation Report" (an appendix to the former report)
>   - — the demonstration test reports produced by the ACWA technology providers and the associated required responses of the providers to questions from the PMACWA
>   - — the PMACWA's demonstration testing database (CD-ROM);
> - perform an in-depth review of the data, analyses, and results of the unit operation demonstration tests contained in the above and update as necessary the committee's 1999 NRC report, *Review and Evaluation of Alternative Technologies for Demilitarization of Assembled Chemical Weapons* (the ACW report);
> - determine if the Burns and Roe, General Atomics, and Parsons-AlliedSignal technologies are viable to proceed with implementation of a pilot-scale program that would employ any of these technologies;
> - produce a supplemental report for delivery to the Program Manager for Assembled Chemical Weapons Assessment.

SCOPE OF THIS STUDY

The committee limited its review to assessing the reports mentioned in the Statement of Task. For each technology package, the committee commented on findings from the initial report that were impacted by the demonstrations (technology-specific findings not related to a demonstrated unit operation are merely noted). This report also includes new findings that may not have been apparent before the demonstration data became available. The committee did not evaluate the extent to which the demonstration tests fulfilled all of the test objectives set by the PMACWA. However, the committee commented on these objectives when they were related to the findings in the initial report (NRC, 1999).

ORGANIZATION OF THIS REPORT

This report consists of five chapters. This chapter has presented background information on the ACWA program and the NRC's involvement in that program. Chapters 2, 3, and 4 discuss the results of the demonstrations for each of the three technology packages. In each chapter, demonstration test objectives are quoted for each unit operation that was demonstrated. (The demonstration objectives are intended to provide contextual technical background [analogous to the Description of the Technology Package sections in the committee's initial report]). Pertinent original findings are discussed, and a concise rationale is given for each of the committee's conclusions on the basis of its review of the documents listed in the Statement of Task. Chapter 5 provides a discussion of the impact of demonstration test results on the original general findings and recommendations. Some new general findings based on the demonstration test results are also provided.

2

Burns and Roe Plasma Arc Process

The plasma arc process proposed by the Burns and Roe team uses modified baseline disassembly for munitions access. Agent, energetics, metal parts, and shredded dunnage are all treated in plasma waste converters (PWCs). The PWCs use plasma arc technology—electrically driven torches with various gases that produce an intense field of radiant energy and high temperature ions and electrons that cause the dissociation of chemical compounds. Materials are processed with steam in the absence of air to produce a plasma converted gas (PCG) that could be used as a synthetic fuel after cleanup and testing.

The integrated PWC system used for the demonstration tests consisted of a PWC—a 300-kW unit capable of operating with a variety of gases (Ar, N_2, CO_2, etc.) in either of two modes: a nontransferred mode (arcing from electrode to electrode on the torch) and a transferred mode (arcing from torch electrode to the melt) (DOD, 1999b). A steam injection system was used for feeding liquids, and a box feed module with a horizontal ram feed was used for feeding solids via a conveyor to the PWC. The gas polishing system, a pollution abatement system, consisted of a quench, a venturi scrubber, a caustic (NaOH) scrubber, a demister, and a high-efficiency particulate air (HEPA) filter.

The PWC system was the only unit operation that was tested. Other components used in the demonstration but not intended to demonstrate a specific unit operation are listed below (DOD, 1999b):

- a liquid feed module
- thermal oxidizers to characterize the effluent from burning PCG
- an energetics deactivation chamber (EDC) for generating and supplying the expected energetics off-gas feed to the PWC

PLASMA WASTE CONVERTER

Demonstration test campaigns of the PWC were planned for treatment of (1) energetics, (2) dunnage and secondary waste, (3) agent, and (4) projectile agent heels.

Energetics Campaign

The energetics campaign was required to validate that the PWC can destroy off-gas from a proposed EDC, which is used for thermal initiation of high explosive components (bursters and fuzes). The following test objectives were established for this campaign (DOD, 1999b):

- Demonstrate the feasibility of the proposed energetics destruction strategy using the integrated EDC demonstration unit and PWC system for high explosives and the PWC system for M28 propellant.
- Validate that the integrated EDC and PWC unit operations can achieve a destruction and removal efficiency (DRE) of 99.999 percent for energetics Comp B and tetrytol.
- Validate that the PWC unit operations can achieve a DRE of 99.999 percent for M28 propellant.
- Characterize the detonation gases and residues from Comp B and tetrytol from the EDC demonstration unit for suitability for processing in the PWC.
- Characterize the deflagration gases from the M28 propellant feed to the PWC system.
- Compare the detonation gases from the EDC demonstration unit to the deflagration gases from the M28 propellant in the PWC system.

The energetics campaign was only designed to show that the PWC could destroy off-gas from the EDC. During the demonstration, 16 grams each of tetrytol and Comp B were detonated in four test runs. Because the design of the detonation chamber was not the one intended for full-scale use, no attempt was made to evaluate its efficacy. Detonation gases were fed to the PWC. (Detonation usually efficiently destroys materials such as tetryl, TNT, and RDX.) The off-gases generated from the EDC were shown to be suitable for feeding to the PWC.

In the opinion of the committee, the use of the EDC would be a poor solution for the destruction of a large volume of energetic materials. During the demonstration tests, M28 propellant was not completely ignited, which was attributed to poor propagation from the initiator. The technology provider explains that initiation at full scale will be accomplished by heating the energetic to 1,100°F. Although a small amount of M28 propellant was introduced directly into the PWC during the demonstration tests, the committee concluded that the test results did not demonstrate conclusively that the direct introduction of propellants would be safe.

Dunnage and Secondary Waste Campaign

The dunnage and secondary waste campaign was required to validate the destruction of solid and liquid secondary wastes and the decontamination of dunnage to a 5X level.[1] Characterization of gaseous, liquid, and solid effluents was required, as was verification of operating parameters. The demonstration tests had the following objectives (DOD, 1999b):

- Demonstrate that the PWC unit operation can process carbon filter media, demilitarization protective ensembles (DPEs), wooden pallets spiked with 4,000 parts per million pentachlorophenol, decontamination solution with carbon filter media, and M55 rocket shipping and firing containers.
- Characterize the process gases, liquids, and solids.
- Validate the ability of the PWC unit operation to meet a 5X condition for solid residues from these feeds.

The demonstration test runs were designed to evaluate the treatment of a variety of dunnage materials, including oak pallets, activated charcoal, fiberglass shipping and firing containers, and DPE materials. Although the test plan originally called for separate testing with each material, the plan was subsequently modified to using a mix of materials. The tests demonstrated the PWC could treat these materials as a mixture, could achieve 5X temperature conditions, and could destroy the pentachlorophenol that had been spiked into the pallets.

The mixed dunnage tests were the only demonstration runs in which sufficient carbon, oxygen, and hydrogen were available in the feed to generate synfuel with appreciable fuel value. The average fuel value of the PCG exceeded 100 Btu/scf in only one of the six mixed dunnage test runs. In several runs, the measurement technique for fuel value failed; in others, the measured average fuel value was very low. In all runs, the oxygen content of the PCG ranged from 5 to 7 percent. This was attributed either to air leakage into the PWC or downstream components or to a lack of control of the oxygen content in the feed materials and gases. The presence of a combustible gas premixed with oxygen clearly represents an unsafe condition susceptible to ignition. Full-scale operation would require design features and/or procedures that would preclude these conditions.

The process did not produce PCG with an acceptable synfuel quality when a steady feed of carbon/hydrogen-containing material was used. Thus, the committee is concerned about the appropriateness, reliability, and robustness of the measurement and control systems. In addition, unless careful control of the steam-to-carbon ratio is maintained, excessive soot may form. Because the system does not include on-line monitoring of the carbon and hydrogen in the feed, the monitoring and control system must reliably measure fuel value and adjust parameters, such as steam flow, to achieve acceptable fuel quality. Such monitoring and control systems were not demonstrated during the test runs, and, therefore, must be developed to ensure the reliable operation of the system with variable feedstocks.

Agent Campaign

The agent campaign was required to validate the destruction of chemical agents. Characterization of gaseous, liquid, and solid effluents was required, as was verification of operating parameters. The test objectives for this campaign are listed below (DOD, 1999b):

- Validate that the PWC process can achieve a DRE of 99.9999 percent for chemical agents HD, GB, and VX.
- Characterize the process gases, liquids, and solids.
- Balance the elemental carbon and heteroatoms from each agent, to the extent possible.

For various reasons, the equipment was not deemed ready for agent tests during the demonstration tests. Therefore, there was no direct demonstration of the ability of the proposed plasma technology to destroy chemical agents. The committee concluded that the variety of equipment problems encountered in the demonstration were due to the immaturity of the proposed integrated process and the particular demonstration equipment, and not due to a fundamental inability of plasma-based technologies to achieve acceptable

[1]Treatment of solids to a 5X decontamination level is accomplished by holding the material at 1,000°F for 15 minutes. This treatment results in completely decontaminated material that can be released for general use or sold to the general public in accordance with applicable federal, state, and local regulations.

results. The history of plasma-based systems for waste treatment indicates that they can destroy chemical agents. Nevertheless, the operability, reliability, and repeatability of the integrated plasma system have not been demonstrated due to equipment failures, system redesigns, and operational modifications. Also, the committee was concerned that some of the agent could bypass the reaction zone (see the discussion below of Finding BR-1 under Review of Previous Committee Findings).

Tests were conducted on the agent-surrogate, dimethyl methyl phosphonate (DMMP), and hydrolysates of HD and VX. In these tests, high DREs of both DMMP and hydrolysate compounds were achieved, increasing the confidence level that the proposed plasma-based process would be capable of destroying chemical agents. However, demonstration tests with neat chemical agents will be required to determine specific operational conditions, such as proper control of oxygen and steam, before pilot-scale evaluations can proceed. These tests will be particularly important for determining the formation of by-products, which is dictated by the materials processed, the stoichiometry for oxygen, steam, and carbon, and temperature conditions. The data on the by-products generated in the demonstration tests are of limited value because the tests were not run with agents.

Projectile Heel Campaign

The projectile heel campaign was required to validate the destruction of chemical agent that had adhered to metal parts and to demonstrate removal of the melt from the PWC. Characterization of gaseous, liquid, and solid effluents was required, as was verification of operating parameters. The test objectives for this campaign are listed below (DOD, 1999b):

- Validate that the PWC process can achieve a DRE of 99.9999 percent for chemical agent GB heels in simulated projectile shells.
- Demonstrate that the PWC can process simulated projectile shell heels using chemical agent in pipe nipples.
- Demonstrate melting of uncontaminated 4.2-inch mortar shells.
- Validate that the PWC unit operation can meet a 5X condition for solid residues from this feed.
- Characterize the gases, liquids, and solids.
- Demonstrate that the melt from the PWC can be removed.

The first five objectives were not met because agent was not injected into the PWC. In addition, the sixth objective was not met because samples were manually removed.

REVIEW OF PREVIOUS COMMITTEE FINDINGS

The committee's earlier findings concerning the Burns and Roe PWC technology package are quoted below and their status following demonstration tests is examined (NRC, 1999):

> **Finding BR-1.** No tests have been done involving actual chemical agent or propellant destruction in a PWC. Tests with agent and M28 propellant were planned for the demonstrations being conducted between February and May of 1999, but no data were available to the committee at the time of this writing.

The demonstration tests conducted on the agent surrogate DMMP (a GB simulant), HD hydrolysate, and VX hydrolysate provided only limited data. The DMMP was 99.99997 percent destroyed; trace levels of thiodiglycol were detected in two of the six HD hydrolysate tests; and the levels of ethyl methyl phosphonic acid and methyl phosphonic acid in the VX hydrolysate tests were very low.

Energetic materials (Comp B and tetrytol) were reported to be 99.9998 percent destroyed, but trace levels of RDX and TNT were detected. Components of M28 propellant were 99.97 percent destroyed (nitrocellulose) and 99.99998 percent destroyed (nitroglycerin). The detection of RDX and TNT in the PWC effluents is indicative that feedstocks can bypass the reaction zone and exit without complete reaction. Thus, if chemical agents were fed to the PWC, they could potentially also bypass the reaction zone and be found in the effluents. Solving this problem will require ensuring thorough mixing in the PWC.

> **Finding BR-2** Scale-up from the small PWC units in existence to the very large units proposed is likely to present significant scientific and engineering challenges.

The numerous problems encountered in the demonstration described above confirmed this finding.

> **Finding BR-3.** Tests performed with one plasma feed gas may not be indicative of PWC performance with a different gas. Because different plasma feed gases have different thermodynamic and chemical properties, the choice of the plasma feed gas could have a significant impact on the performance of the system. For example, the electrical power requirements will be determined, in part, by the plasma feed gas. Electrode wear may also depend on the type of gas, and product gas composition will vary.

Initially, the technology package proposal indicated that argon would be used as the plasma feed gas. This would distinguish the PWC from an incinerator because the inert gas is not an oxidizing agent. Citing the expense of argon, the technology provider subsequently shifted to carbon dioxide (CO_2), which is cheaper, but introduces a source of oxygen. Computer calculations for various chemical agents introduced into a CO_2 plasma at ~ 3,000 K predicted that agents would undoubtedly be destroyed but also indicated that large amounts of carbon soot would be formed as the hot gaseous mixture cooled. The presence of particulates of high surface area (that are probably pyrophoric) in the product creates a new problem. Also, electrical power requirements for CO_2-plasma operation would be greater than for argon-plasma operation.

In the actual demonstration tests, nitrogen (N_2) was used as the plasma gas. Although N_2 is a nonoxidizing species, reaction products of environmental concern (C_2N_2, HCN, metal cyanides, etc.) were predicted and were detected in the demonstration tests. The power requirements for N_2-plasmas are acceptable.

In summary, the technology provider has explored a few alternatives for plasma gases but may not have found the best choice. Also, the problem of torch failure could be minimized by a better choice of metals or by alternative designs. For water-cooled plasma torches, the metals must not react with the plasma gases and must still have high melting points to prevent a sudden release of water into the PWC (see the discussion following Finding BR-5).

> **Finding BR-4.** The technology provider's proposal for recycling the liquid-scrubber effluent through the PWC to vitrify the salts may not be practical. If scrubber liquor is fed to a PWC, some of the contaminants may simply revolatilize. In addition, NaCl and NaF salts could react with SiO_2 at high temperatures to form gaseous $SiCl_4$ and SiF_4, respectively (both hazardous materials).

The demonstration tests did not address the ability of the PWC to vitrify salts from recycled scrubber liquor. Finding BR-4 remains unchanged.

> **Finding BR-5.** The maintenance of negative pressure within the PWC has not been demonstrated under munition-processing conditions. Pressure excursions that produce positive pressure in the PWC vessel could release product gas to the surrounding room. Some upsets that could result in moderate to severe pressure excursions included:
>
> - A leak in the torch-cooling system to release water into the PWC, and rapid steam formation could pressurize the vessel.
> - Energetic material that remained in a mortar or projectile introduced into a PWC could detonate upon heating, which would generate a pressure pulse.
> - An improper cut of the rocket motor could allow a larger-than-design piece of propellant to be introduced into the PWC. If the gas production rate from the propellant exceeds the capacity of the downstream PAS, the vessel could overpressurize.

The primary safety problem apparent from the demonstration tests is an inability to maintain negative pressure. Overpressurization occurred several times during the tests due both to plasma torch failure and poor engineering system design (e.g., ram feeder blow-back and leaks in the gas polishing system). The failure of the plasma torch caused cooling water to be released into the PWC, which could have resulted in catastrophic overpressure that could have released agent, if any had been present. Thus, substantial further engineering development will be necessary, along with design and administrative controls to ensure the safe use of this plasma torch technology.

According to the technology provider's proposal, rocket propellant would be sent directly to the PWC, whereas explosives would be sent first to the EDC. Although a small amount of the propellant was tested in the PWC, the committee was concerned that larger amounts of propellant might detonate rather than deflagrate. The resolution of this issue has not been successfully demonstrated.

> **Finding BR-6.** Combustion of plasma-converted gas in a boiler faces three major hurdles: (1) to avoid being permitted under RCRA as a boiler burning hazardous wastes, the gas may have to be delisted; (2) the gas may require significant scrubbing to remove compounds that are unsuitable as boiler feedstock; and (3) the boiler will have to be configured to burn gas that has a low heating value efficiently in order to avoid generating unacceptable emissions.

The Environmental Protection Agency (EPA) has recently established an exemption for synfuel produced from hazardous waste. Under the Comparable/Syngas Fuel Exclusion (40 CFR 261.38), synfuels that meet certain specifications are not classified as hazardous wastes and, therefore, could be burned without Resource Conservation and Recovery Act (RCRA) permits in boilers and industrial furnaces (a Clean Air Act [CAA] permit would still be necessary). The synthesis gas fuel specification has the following criteria:

- a minimum Btu value of 100 Btu/scf
- less than 1 ppm_v of total halogen
- less than 300 ppm_v of total nitrogen other than diatomic nitrogen (N_2)
- less than 200 ppm_v of hydrogen sulfide
- less than 1 ppm_v of each hazardous constituent on a target list of 40 CFR 261 Appendix VIII constituents

These stringent requirements were not met in any of the demonstration tests. It was not clear that the tests were designed to evaluate this specification, even though it would be critical to the development of an alternative disposal technology using PCG. Without this exemption, the PCG synfuel could not be used in boilers without a RCRA/CAA hazardous waste combustor permit subject to boiler and industrial furnace rules (the so-called "BIF rules").

The demonstration tests revealed several potential problems with PCG meeting the Comparable/Syngas Fuel Exclusion. Only one material tested in the demonstration (mixed dunnage) was converted to synfuel with an appreciable fuel value. Even for this material, the minimum Btu value (> 100 Btu/scf) was only demonstrated in one test (out of six). For all other tested materials, the Btu value of the synfuel was very low (generally close to zero).

Furthermore, both the generation of hazardous air emissions and the conversion of carbon are strongly affected by carbon/oxygen stoichiometry. The generation of synfuel of insignificant Btu value in nearly all of the demonstration test runs casts doubt on the relevance of the emissions data to full-scale operation for most of the materials tested in the demonstration. The Comparable/Syngas Fuel Exclusion specification for hazardous constituents would have to be demonstrated for the specific conditions that would yield a PCG with acceptable Btu value. In addition, a more complete profile of all 40 CFR 261 Appendix VIII compounds would have to be evaluated. Finally, the demonstration tests did not confirm that predictable control of the PCG heat content could be achieved, even when higher hydrocarbon feedstocks (such as mixed dunnage) are treated.

Another challenge to meeting the Comparable/Syngas Fuel Exclusion requirements for PCG is maintaining low levels of nitrogen and halogen compounds. The NO_x-level for every PCG material tested in the demonstration unit exceeded the criterion of less than 300 ppm_v (~388 mg/m^3) of total nitrogen other than diatomic nitrogen (N_2). The removal of nitrogen compounds from the PCG synfuel is a critical process capability that was not demonstrated but is necessary for full-scale operation.

The PCG synfuel generated from all of the test materials (except tetrytol) exceeded the total halogen requirement of 1 ppm_v (~1,500 μg/m^3) for chlorine despite the use of an acid scrubber. For example, PCG generated from DMMP in the demonstration tests had 26,980 μg/m^3 (approximately 18 ppm_v) of chlorine, which is 18 times the Comparable / Syngas Fuel Exclusion of 1 ppm total halogens. Thus, the demonstration unit also failed to demonstrate that it could generate synfuel that meets these critical synfuel exclusion criteria.

> **Finding BR-7.** Although a PWC may not be considered to be an incinerator by permitting authorities, the most likely permitting procedures for a PWC would be similar to those used for incinerators.

A key component of the Burns and Roe demonstration tests was to determine the characteristics of the flue gas when the synfuel is burned in the thermal oxidizer. These characteristics can suggest the emissions from a boiler or industrial furnace burning the PCG. In other permitting actions relating to plasma units that generate gas burned in catalytic oxidizers (e.g., the ATG facility in Richland, Washington, EPA Region 10), the EPA and state regulators used appropriate, relevant, and applicable rules (ARARs) based on the hazardous waste combustion rules.

A comparison of the thermal oxidizer emission levels with the Hazardous Waste Combustion ARARs indicates that either additional cleanup of the PCG would be required or the emissions of the boiler/industrial furnace would require more rigorous scrubbing. This comparison is complicated by the highly dilute conditions in some of the thermal oxidizer exhaust (i.e., 12 to 20 percent oxygen). It is also worth noting that the thermal oxidizer used would not generally meet the carbon monoxide standard of 100 ppm_v. A comparison of the hazardous waste combustion rules with the thermal oxidizer emissions data indicates that the combustion of PCG would not meet some standards, when corrected to the standard 7 percent oxygen, (e.g., the cadmium-plus-lead emission for the system configuration used in the demonstration tests for M28 propellants, mixed dunnage, and VX hydrolysate). Mercury emission could be a problem for M28 propellants, and particulate matter would be a problem for the treatment of mixed dunnage. Chlorinated dioxin/furan was not found to be problematic for the configuration demonstrated when compared to the hazardous waste combustion standard. In summary, the demonstration tests did not show that the PWC system could adequately control emissions for the direct combustion of PCG in a boiler or industrial furnace.

SAFETY ISSUES

In the earlier report, the committee made the following observation (NRC, 1999):

> Cooling water is circulated through the plasma torch to keep it from melting at the high plasma temperatures. A leak in the cooling system could spray water into the plasma. If the leak is sudden, rapid vaporization could cause a pressure pulse that might overload the downstream gas-handling equipment. Then, untreated agent could be released into the surrounding room through the torch opening in the top of the PWC. Similar "puffing" has been observed in combustion equipment when excessive back pressure occurs. If the leak is gradual, the resulting steam would dissociate in the plasma forming hydrogen and oxygen gas that could recombine and explode if the mixture is in the flammable range above its autoignition temperature. The effect of liquid water introduced into a plasma in the presence of other species present in PWCs must be determined before larger scale experiments are performed. . . .
>
> The technology provider is aware that torch failure is a concern, and the potential for an explosion has been reduced by the torch design and by redundant flow and pressure controls that would actuate fast-closing valves on the water feed as well as the waste feed in the event of a failure.

The committee reiterates its earlier observation that appropriate design and administrative controls can ensure the safety of plasma arc technology (NRC, 1999).

The technology provider proposes sending rocket propellant directly to the PWC, whereas, explosives will be sent first to the EDC. Although a small amount of the propellant was demonstrated to deflagrate in the PWC, the committee is concerned that larger amounts of propellant may detonate rather than deflagrate. The committee does not believe this issue has been successfully demonstrated.

The addition of nickel to the melt to form a conductive bed for the transferred arc operation constitutes another issue regarding worker safety (Burns and Roe, 1999a). Airborne nickel particulate is very hazardous and should be assessed further with respect to worker exposure during normal operations, anticipated transient conditions, maintenance, and accidents.

The recovery of molten metal may require more access

by workers during operations, as well as increased maintenance. Increased access would also increase worker exposure to hazards over predemonstration estimates.

REEVALUATION OF STEPS REQUIRED FOR IMPLEMENTATION

The committee's earlier report identified the following five steps required for implementation (NRC, 1999):

1. Determine the effect of sudden water injection into the plasma torch in the presence of argon, nitrogen, carbon dioxide, and other species present in the plasma system. Include an evaluation of the effect of gases present in the PWC on the flammability range of hydrogen gas.

2. Determine the likelihood of the release of untreated agent and other hazardous contaminants from the PWC if the gas generation rate is unexpectedly high (e.g., due to a cooling-water leak, the inadvertent introduction of explosive material into the chamber, or a rapid deflagration of propellant).

3. Conduct a thorough analysis of the product gas generated from each PWC using the plasma feed gas proposed for full-scale operation. This analysis should include the identification of organic intermediates that would be of concern in an HRA [health risk assessment].

4. Establish the efficacy of pollution-control equipment in removing hazardous compounds (e.g., NO_x, SO_x, HCl, and metals) from the product gas.

5. Perform a larger-scale demonstration of PWC operation that includes the hold-test-release step.

None of these steps was completed in the demonstration tests. Furthermore, the test results do not readily indicate how the concerns raised by the committee could be addressed.

Clearly, extensive testing with chemical agents will be necessary if PWCs as currently proposed by the technology provider are to be used. As discussed in Finding BR-5, serious doubts have been raised about the reliability of the torch design and the maintenance of negative pressure in the system, and, hence, about the safety/efficacy of this system.

The committee believes a properly configured and operated plasma arc process would be a robust, indiscriminant thermal process capable of destroying chemical agents. However, on the basis of observations during two site visits to plasma arc installations (Ontario Hydro Technologies, Toronto, Ontario, and Aberdeen Providing Ground, Maryland), the results of the demonstration tests, and a review of the available demonstration data, the committee concurs with the Army's conclusion that the Burns and Roe process is too immature to be considered as a viable solution for the destruction of assembled chemical weapons at this time.

SUPPLEMENTAL FINDINGS

Finding BR-1. The plasma torch apparatus, as demonstrated by the Burns and Roe team, is not qualified for further consideration for the demilitarization of assembled chemical weapons. The torch design appears to be unreliable for extended use. Furthermore, the design increases the possibility of a catastrophic water leak, which could produce a significant increase in pressure in the PWC, and possibly cause an explosion, which, in turn, could expose personnel to chemical agent. Moreover, the effectiveness of the monitoring and control sensors was not demonstrated.

Finding BR-2. Even after more than a year of research and development, the technology provider has not been able to show that its small PWC can adequately destroy agent simulants or that nitrogen is the best gas to use for the plasma feed. If oxygen leaks into the reactor, it could react violently with hydrogen. If air were used for the plasma feed gas, regulatory compliance issues would arise, as well as questions of public acceptance.

Finding BR-3. In the absence of any data for processing effluents from agent runs, the committee could not validate the ability of the proposed system to handle and stabilize effluent products from agent processing.

3

General Atomics Technology Package

The General Atomics process uses a modified version of the baseline disassembly process and cryofracture of projectiles for munitions access. The agent and energetics are destroyed by hydrolysis. The hydrolysate is then treated by supercritical water oxidation (SCWO). Metal parts are subjected to caustic hydrolysis processing followed by 5X thermal treatment. Dunnage is shredded, mixed with caustic, and destroyed by SCWO.

Demonstration tests were conducted for the following operations:

- energetics rotary hydrolyzer (ERH)
- dunnage shredding and hydropulping
- SCWO

ENERGETICS ROTARY HYDROLYZER

The objectives of the demonstration tests of the ERH are listed below (DOD, 1999b):

- Demonstrate the effective dissolution of aluminum and energetics in fuzes and bursters, as well as propellant in rocket motors, to allow downstream processing in the continuously stirred tank reactor, SCWO reactor, and heated discharge conveyor.
- Determine the deactivation of the energetics in fuzes and bursters and the propellant in rocket motors.
- Validate the retention times for aluminum and energetics in fuzes and bursters and propellant in rocket motors
- Characterize the gas, liquid, and solid process streams.

The General Atomics demonstration tests involved several different munition items and energetic materials. Complete destruction, (i.e., below the detection limit) was achieved for tetryl in M557 fuzes and M14 bursters and for tetrytol (tetryl/TNT) in M6 bursters. However, the following problems arose during the handling of other energetics (General Atomics, 1999a):

- Small quantities of fuze-train components remained unhydrolyzed; these were destroyed in the hot muffle furnace.
- Unhydrolyzed energetic material adhered to a flight drum during an M83 burster (RDX/TNT) validation test and burst into flame.[1] (The technology provider claims that this was an artifact of the test; the flights in the ERH were designed to hold solids and liquids for sampling rather than to drop them into the hydrolyzing solution. An appropriate flight design will be used in the full-scale ERH).
- Excessive boiling and foaming was reported with the M83 burster, which could cause difficulties in processing.
- RDX and HMX were above the detection limit in the liquid analyte.
- Hydrolysis of M28 propellant in the motor casing was slower than anticipated; the NaOH solution concentration had to be raised to 12M. (The technology provider has suggested cutting the propellant into smaller pieces).
- During the processing of M28 rocket propellant, a yellow substance (identified as N-nitrosodiphenylamine) was generated and coated much of the interior of the explosive containment cubicle. The technology provider indicated that the coating was caused by the ventilation flow in that particular ERH test unit. The ventilation was sized to dilute hydrogen to below the lower explosive limit and was clearly inadequate to prevent

[1]The term "flight" refers to plates attached to the drum that hold the energetic pieces as the drum rotates. The entire apparatus is called a flight drum.

fugitive emissions from the ERH. The technology provider reported that in the full-scale system, sufficient ventilation flow would be provided to prevent fugitive emissions (General Atomics, 1999b). The yellow material would be scrubbed from the ERH ventilation flow, and the scrubber solution would be combined with energetics hydrolysate and processed through the SCWO reactor. The committee was concerned that the proposed solution to the problem could result in the accumulation of similar energetic by-products in other parts of a full-scale system.

The committee's earlier report contained the following finding concerning hydrolysis of energetics (NRC, 1999):

> **Finding GA-2.** Hydrolysis of energetics at the scales proposed by the technology provider is a relatively new operation. Chemically, it is possible to hydrolyze all of the energetic materials; however, the rate of hydrolysis is limited by the surface area and, therefore, depends on particle size. (Smaller particles are more desirable because they have a higher surface-to-volume ratio.) The proposed method of removing and hydrolyzing the energetics appears to be reasonable, but further testing is required to determine the hydrolysis rates and to confirm that throughput rates can be achieved.

The demonstration tests substantively confirmed this finding. The test results demonstrated that the ERH could deactivate and dissolve the energetics and aluminum found in M557 fuzes and M83 bursters and could deactivate the energetics found in M6 and M14 bursters in two to four hours. Test data on the M28 rocket motor sections show that a residence time of 10 hours at 12M caustic concentration and 230°F were required for complete hydrolysis of the M28 propellant.

The demonstration program did not include the treatment of agent-contaminated solids. In the opinion of the committee, the ability of the ERH system to hydrolyze solid pieces of propellant supports the conclusion that similar treatment could successfully clean contaminated solids to a 3X condition.[2] However, the demonstration results cannot be considered conclusive evidence that the required processing rates could be consistently achieved.

The committee's earlier report included the following finding (NRC, 1999):

> **Finding GA-3.** The rotary hydrolyzer appears to be a mature reactor configuration that is well suited for this application.

Although no test data on the reaction rate were provided, the tests did qualitatively demonstrate that the ERH could destroy energetic materials. However, some results indicate that the ERH did not completely wet the energetics with hydrolysis solution, which allowed some solid energetic material to exit the ERH before hydrolysis was complete. The explanation given by the technology provider (i.e., the shape of the flights) and the design modification proposed by the provider to address this problem (i.e., modification of the pitch and shape of the flights) should, in the committee's opinion, decrease the amount of unexposed solid material that passes through the ERH.

No tests were conducted on the hydrolysis of energetics contaminated with agent; however, because of the long residence time in the ERH, the committee believes that chemical agent exposed to the caustic hydrolysis solution in the ERH would be hydrolyzed. Nevertheless, because the exact manner in which agent might penetrate energetic materials is not known, there is still some question as to whether chunks of unhydrolyzed energetic material, such as those that were found in the residue from the ERH, would be truly agent free. Agent embedded in the energetic solids might not have been exposed to the caustic solution and, hence, might not have reacted.

DUNNAGE SHREDDING/HYDROPULPING SYSTEM

The purpose of the demonstration tests of the dunnage shredding/hydropulping system (DSHS) was to show that solid wastes (wooden dunnage, DPE suits, and butyl rubber) could be adequately reduced in size and pulped to a pumpable mixture. The objectives of the demonstration testing are listed below (DOD, 1999b):

- Validate that the shredders and hydropulper can adequately prepare the dunnage for downstream processing in the SCWO reactor.
- Qualitatively evaluate the operability (especially material handling) of the shredder/hydropulper unit operations.
- Validate that the shredders can process 1,000 lb/hr of pallets and, separately, 250 lb/hr of plastics.

Several commercial shredders identical in size to the units proposed for the full-scale system were used to achieve the size reduction of the solid materials of interest. In the initial report, the committee had stated the following (NRC, 1999):

> **Finding GA-4.** Shredding of dunnage and injection of the slurry directly into a SCWO system is a new and unproven process. While General Atomics claims to have developed a proprietary pump capable of pumping the slurry at high pressures, it has not been tested under the

[2]At the 3X decontamination level, solids are decontaminated to the point that agent concentration in the headspace above the encapsulated solid does not exceed the health-based, eight-hour, time-weighted average limit for worker exposure. The levels for HD, VX, and GB are, respectively, 3.0, 0.01, and 0.1 µg per cubic meter in air. Materials classified as 3X may be handled by qualified plant workers using appropriate procedures but are not releasable to the environment or for general public reuse. In specific cases in which approval has been granted, a 3X material may be shipped to an approved hazardous waste treatment facility for disposal in a landfill or for further treatment.

intense solids loading anticipated. Furthermore, the injection of large amounts of solid material, including wood shreds, cut-up nails, and complex organic materials, such as pentachlorophenol and other wood preservatives, into the SCWO system has not been demonstrated. Considering the difficulty SCWO reactors have encountered with deposition of solids when liquids are treated, the committee believes that this application of SCWO may encounter significant difficulties. (At the time of this writing, processing of solids with SCWO was being performed as part of the ACWA demonstrations.)

The individual components of the DSHS had been tested previously in their respective applications but had not been used collectively in the configuration used for the demonstration test program. Consequently, numerous, albeit surmountable, problems were encountered (e.g., wood "nesting" in the hammer mill and micronizer feed chutes and inadequate magnetic separation of metal from the shredded DPE suits prior to processing in the granulator). The technology provider was able to control both system and feed variables well enough to achieve the targeted feed processing rates and obtain the proposed objective for size reduction (< 1 mm for wood and < 3 mm for plastics). The 3-mm plastic material product was processed through a sieve to separate material that was less than 1 mm that could be fed to the SCWO reactor. The full-scale SCWO system will have larger feed nozzle diameters that should be capable of accepting the plastic dunnage material as shredded (i.e., without the need for sizing to less than 1 mm) (General Atomics, 1999b).

The demonstration tests did not validate that the hydropulper could consistently produce material that was smaller in diameter than the goal objective of 1-mm; however, the tests did determine that the hydropulper could blend energetics hydrolysates with size-reduced wood to yield a uniform, pumpable slurry for processing in the SCWO reactor.

The mass balance reported for the two validation test runs of the micronizer while processing wood pallets showed a 5.4 and 6.3 percent deficit (General Atomics, 1999a). The deficit was attributed to "Presumably . . . the loss of water due to heatup in the micronizer." This loss is not a problem for pallets that are not contaminated with agent. When contaminated wood is processed, however, the water vapor released could contain vaporized agent, and the gas stream will have to be managed accordingly.

The duration of the shredding tests was too short to allow for an evaluation of the long-term efficacy of this process. The demonstration was highly labor intensive and, because it was performed on uncontaminated material, did not require that the operators work in full protective clothing. Therefore, it cannot be concluded that a full-scale system would provide similar levels of materials segregation without further development of the process. For example, one of the technology provider's conclusions is that the metal parts in DPE suits would have to be manually cut out in glove boxes prior to processing and then decontaminated to a 5X condition in the metal parts furnace (General Atomics, 1999a). Because this step (which is necessary for successful processing) was not performed during the demonstration tests, the committee could not assess its efficacy.

SUPERCRITICAL WATER OXIDATION SYSTEM

The hydrolysates of energetic materials provided by the PMACWA were prepared using 12 percent sodium hydroxide (as specified in the technology provider's proposal). The DREs from hydrolysis of energetic constituents of Comp B, tetrytol, and M28 propellant all exceeded 99.999 percent, except for the nitrocellulose component of the M28 propellant. The latter was set to measure a DRE of no greater than 99.988 because the analytical method has a high detection limit.

The SCWO system was demonstrated to validate its capability to destroy Schedule 2 and other organic compounds produced from agent hydrolysis. The objectives of the demonstration testing of the SCWO reactor concerning agent hydrolysate products are listed below (DOD, 1999b):

- Validate that the SCWO reactor can eliminate the Schedule 2 compounds present in the agent hydrolysate feed.
- Validate that the agent hydrolysis process and the SCWO reactor can achieve a DRE of 99.9999 percent for HD, GB, and VX.
- Demonstrate the long-term operability of the SCWO reactor with respect to salt plugging and corrosion.
- Characterize the gas, liquid, and solid process streams from the SCWO reactor.

The SCWO system was also demonstrated to validate its capability to destroy organic compounds from energetic hydrolysis products and to demonstrate the feasibility of destroying shredded dunnage. The demonstration tests included the following objectives (DOD, 1999b):

- Validate that the ERH, continuously stirred tank reactor (CSTR), and SCWO can achieve a DRE of 99.999 percent for tetrytol, Comp B, and M28 propellant.
- Determine the impact of the aluminum from the ERH process on SCWO operation.
- Determine how well organics in the shredded dunnage are oxidized in the SCWO reactor.
- Characterize the gas, liquid, and solid process streams from the SCWO system.

The committee's initial report contained the following finding concerning General Atomic's use of SCWO (NRC, 1999):

Finding GA-5. All of the findings in the [1998] NRC report, *Using Supercritical Water Oxidation to Treat*

Hydrolysate from VX Neutralization, apply to the General Atomics system.

The demonstration confirmed this finding (see Appendix A). Although the SCWO system successfully destroyed organic compounds in the liquids, the results did not demonstrate that the system is capable of operating without frequent shutdowns for repair or cleaning. This uncertainty could affect the system's ability to treat the numbers of munitions located at a storage site within a reasonable length of time. For the destruction of agent and energetics hydrolysates and dunnage, the SCWO system performed reasonably well. However, corrosion and salt plugging both raised concerns about reliable long-term operation.

Operationally, the validation test runs for agent hydrolysate (all liquid feeds) proceeded smoothly, except for inconsequential leaks at some joints. Validation test runs for the energetic hydrolysates and dunnage feeds showed that these can be processed successfully, provided that aluminum hydroxide is removed from the feed (it caused severe plugging). Safety issues pertaining to the removal of aluminum hydroxide are noted later in the chapter.

Thus, the demonstration confirmed the concerns of the committee (and of another NRC committee that had previously evaluated the use of SCWO to treat VX hydrolysate) about the durability of components and the materials of construction in the highly corrosive SCWO system environment (NRC, 1998, 1999). Although the demonstration plan had called for the use of a platinum-lined reactor, because of problems encountered in fabricating the platinum liner, an unlined Inconel™ 718 SCWO reactor was used. This contributed to the corrosion and plugging of the downstream components with corrosion products (DOD, 1999b).

SCWO processing of the dunnage slurry was not demonstrated beyond a simple proof of concept. As described in the technology provider's report, a mixture of tetryl hydrolysate, aluminum hydrolysate, deionized water, phosphoric acid, micronized wood, granulated plastic (< 1 mm), ground activated carbon, and a stabilizing additive proprietary to the technology provider was fed to the SCWO reactor at an approximate rate of 6 kg/hr (General Atomics, 1999a). The committee concluded that this brief test constituted a proof of concept only and could not be considered a validation of the method.

The demonstrated treatment of shredded and slurried dunnage using SCWO resolved one of the committee's concerns but raised new ones. The demonstration tests showed that the SCWO system's pump can pressurize the slurry to the high pressure required for the SCWO reactor and that the SCWO reactor is capable of oxidizing the slurried dunnage. However, the testing did not demonstrate that tramp metal[3] would not prove to be a problem in extended operation. Furthermore, the demonstration tests of the SCWO system with dunnage feed was too short to demonstrate the long-term reliability of the system.

Finally, the demonstration tests used slurried solids of dunnage shredded to less than 1 mm (rather than less than 3 mm as proposed in the full-scale process), and the feed nozzles were smaller than those proposed for full-scale operation. Thus, the efficacy of the process with particles sized to full-scale specifications and larger nozzles was not demonstrated.

SAFETY CONCERNS

The demonstration tests revealed that additional processing steps to remove aluminum from energetics hydrolysate would be necessary to prevent plugging of the SCWO reactor. The technology provider has proposed using a neutralization and filtration process to remove aluminum hydroxide from the hydrolysate, with subsequent 5X treatment of the precipitated aluminum filter cake in an inductively heated metal parts furnace (General Atomics, 1999b). Aluminum hydroxide forms a very flocculent precipitate, however. Because this compound is also amphoteric, the pH will have to be carefully controlled and the precipitate carefully filtered. If other hazardous metal salts precipitate with the aluminum hydroxide, they may have to be treated under RCRA specifications.

The removal of aluminum hydroxide would require additional processing equipment, which would add to the maintenance and reliability burden of the plant and would increase worker maintenance time in DPE suits and opportunities for worker exposure to agent. This concern was raised in the committee's initial report (NRC, 1999). It is repeated here to emphasize that modifications used in the demonstration tests would increase the potential of exposure.

The demonstration tests showed that condensable organics, such as nitroglycerine, will be evolved from the ERH and will be subsequently condensed and returned to the CSTR for hydrolysis. The committee notes that considerable care will be required to ensure that these condensable explosive materials are not initiated, thereby increasing the possibility of worker exposure to agent and damage to process equipment. The ERH demonstration tests using propellant feed also resulted in the release of volatile organic compounds (VOCs) into the explosive containment cubicle for the ERH. The walls of the cubicle were coated with this material as it condensed. As the technology provider noted, this experience reveals that the ERH design will have to control fugitive emissions (General Atomics, 1999b). The committee believes that the potential for worker exposure to agent would be increased during the maintenance of currently undefined control systems for fugitive emissions.

The technology provider also indicated that, to preclude dust explosions (which are extremely unlikely) in the micronizer component of the DSHS, additional safety

[3]In this instance, tramp metal consists of metal pieces and fragments originating from dunnage components entrained in the dunnage slurry.

features will be required for the full-scale design of the system (General Atomics, 1999b). These features, too, would increase the opportunities for worker exposure to agent during maintenance.

In general, the demonstration tests revealed that more maintenance in DPE suits would be required, and, thus, the opportunities of exposure to agent by workers would be increased. At the baseline incineration disposal facilities operating on Johnston Island (in the Pacific Ocean) and at Tooele, Utah, workers in DPE suits are only allowed to remain in contaminated areas for two hours at a time and can only enter if another worker is present. In case of emergency, two more workers wearing protective clothing must be prepared to provide assistance (PMCD, 1998). Thus, an increase in maintenance in DPE suits can have a significant impact on productivity. Process design and the selection of reliable process equipment and materials, in conjunction with suitable training and procedures, should be used to minimize requirements for activities in DPE suits.

EFFLUENT CHARACTERIZATION

In the initial report, the committee concluded that the liquid effluent from the General Atomics process consists of pure water from the evaporator/crystallizer used to produce the solid filter cake (NRC, 1999). This effluent is essentially distilled water and should not pose a significant hazard to human health or the environment. The solid waste from the process, consisting of dried filter cake, was reasonably well characterized. The gaseous effluent from the SCWO process was not well characterized, however, and as a result, its hazardous characteristics could not be determined.

Tables 3.4-10 through 3.4-19 in the demonstration test report by General Atomics present some analytical results on the liquid and gaseous effluents from the SCWO reactors (General Atomics, 1999a). However, the reported characterizations are inadequate to determine if the solid filter cake could be stabilized adequately or to estimate the degree of risk to human health or the environment posed by the gaseous effluent from the SCWO process.

A further concern relates to the presence of sodium and other solid materials in the gaseous emissions from the SCWO reactor. The mechanism whereby solids are released into the gaseous effluents is not clear. One would expect that these inorganic materials would be found in the solid and liquid phases, but not in the gaseous phase. Small quantities of chromium in the gaseous emissions from the SCWO reactor are of potential concern for two reasons. First, it reinforces the importance of demonstrating the reliable operation of the platinum-lined reactor; second, it illustrates the need to test gaseous emissions from the SCWO system for particulates, as well as for gaseous contaminants. Chromium emissions reported in Table 3.5-11 of the technology provider's demonstration test report were at 3.1, 12.3. and 10.5 micrograms, respectively, during a five-hour test period for each of three test runs (General Atomics, 1999a). If the reported emissions pertain to chromium in the hexavalent form, the committee has serious concerns.

Table 3.4-8 of the demonstration test report by General Atomics shows that chemical analyses on VOCs and semi-VOCs were conducted on samples from the off-gas duct of the SCWO system during tests with HD hydrolysate tests (General Atomics, 1999a). The results of these measurements, however, do not appear to be adequate for evaluating the environmental impact of the process. Standard EPA methods for analysis of gaseous effluent samples generally produce full scans that can indicate the quantities of a large number of compounds of environmental concern.[4] These results, along with the results for emissions of metals (including chromium valency), can then be used to assess the environmental impact of a facility through accepted risk assessment methods (EPA, 1998a).

REEVALUATION OF STEPS REQUIRED FOR IMPLEMENTATION

The committee's earlier report included six required steps for implementation of the General Atomics overall technology package (NRC, 1999). These steps are reprinted below, followed by a description of the effects of the demonstration tests on them.

1. Conduct tests of the cryofracture process to ascertain if it provides better access to the agent cavity in projectiles and mortars then the baseline disassembly process.

Cryofracture was not part of the demonstration.

2. Sample and analyze air emissions from the demonstration system. The air emissions will have to be measured to a level of detail and accuracy that can be used for HRAs and environmental risk assessments required by EPA (1998a).

Some sampling and analyses of air emissions were conducted during the demonstration. However, additional data will be required to evaluate HRA and EPA emissions requirements.

3. Verify that energetic materials encased in metal (e.g., rocket or other munitions fragments) will be hydrolyzed.

The demonstration tests did verify that energetic materials encased in metal can be hydrolyzed. They also confirmed that the chemical reaction of the aluminum casings with the caustic solution is sufficient to gain access to and hydrolyze the contained energetic materials in the design residence time of the ERH.

---

[4]8000 Series Methods, especially those using gas chromatography/mass spectrometry scans (e.g., Methods 8260B, "VOCs by GC/MS," and 8270C, "Semi-VOCs by GC/MS").

4. Ascertain how well the SCWO process can handle high-solids materials (shredded dunnage).

The demonstration indicated that the SCWO process can handle materials with a high solids content (e.g., shredded dunnage). However, the SCWO system was not operated long enough to demonstrate reliable continuous operation.

5. Ascertain how well the SCWO system can treat hydrolysate containing large amounts of chlorides, sulfur, and phosphates on a continuing basis.

The ability of the SCWO system to treat hydrolysate containing large amounts of chlorides, sulfur, and phosphates on a continuous basis was not demonstrated.

6. Determine erosion and corrosion behavior of the components of the SCWO system.

General Atomics provided data on the types and quantities of metals found in the precipitates. Both the types and relative quantities matched those of Inconel™ 718. These data provide a strong indication that Inconel™ 718 was the source of the precipitates during the demonstration tests; they do not prove that other materials would not also form precipitates. In addition, the results do not confirm that a platinum-lined reactor could withstand the SCWO conditions and protect the underlying reactor wall during sustained operation.

SUPPLEMENTAL FINDINGS AND RECOMMENDATIONS

Finding GA-1. Testing on the hydrolysis of energetic materials contaminated with agent will be necessary before a full-scale system is built and operated.

Finding GA-2. Testing will be required to verify that the larger diameter supercritical water oxidation (SCWO) reactor feed nozzles will be capable of accepting the dunnage material as shredded (i.e., without additional classification and segregation) and that the reactor will perform reliably under these conditions.

Recommendation GA-1. Operation of the size reduction and slurrying system, and long-term operation of the supercritical water oxidation (SCWO) reactor with slurry, should be conducted before proceeding with a full-scale system.

Recommendation GA-2. Before construction of a full-scale supercritical water oxidation (SCWO) system, additional evaluations of construction materials and fabrication techniques will be necessary because corrosion and plugging prevent continuous operation with the present design. If the new construction materials do not solve these problems, then alternative SCWO reactor designs should be investigated.

Recommendation GA-3. To determine the operability of the supercritical water oxidation (SCWO) reactor and the reliability of the materials of construction, long duration runs of a SCWO reactor should be conducted with slurry, with energetics hydrolysate, and with agent hydrolysate before full-scale implementation proceeds.

Recommendation GA-4. The efficacy and safety of the additional step to remove aluminum hydroxide from the hydrolysate produced from rocket propellants should be evaluated prior to construction of a full-scale supercritical water oxidation (SCWO) system.

Recommendation GA-5. Decontamination of solid munitions materials by flushing and immersion should be demonstrated prior to full-scale implementation.

Recommendation GA-6. The air emissions data from the demonstration tests should be used in a screening risk assessment. The results of the air effluent samples should be subject to (1) a human health risk assessment following the Human Health Risk Assessment Protocol (HHRAP) for Hazardous Waste Combustion Facilities from the Environmental Protection Agency (EPA) [EPA530-D-98-001(A,B,C)], and (2) an ecological risk assessment following a protocol that will be released by EPA in the very near future.

4

Parsons-AlliedSignal WHEAT[1] Process

The Parsons-AlliedSignal process uses a modified baseline disassembly system to access agent and energetics. These are then hydrolyzed and subsequently biotreated. Hardware and dunnage are thermally decontaminated. The overall process consists of several technologies.

Demonstration testing was conducted for the following components:

- munitions cutting and fluid mining
- biotreatment of agent and energetics hydrolysates
- catalytic oxidation of all gases from the process
- metal parts treater for 5X treatment and dunnage treatment

MUNITIONS CUTTING AND FLUID MINING

A major modification from the standard baseline disassembly system is the use of water-jet cutting for accessing munitions, followed by fluid mining (wash out) of the energetics with high-pressure water jets. The objectives of the tests are listed below (DOD, 1999b):

- Demonstrate that circumferential cuts at required locations along the rocket length can be made.
- Demonstrate effective fluid mining and separate collection of rocket bursters, motor propellants, and residual agent simulant.
- Demonstrate that control can be maintained of rocket metal and plastic parts from cutting and fluid mining operations.
- Determine the energetic particle size of mined rocket bursters and propellant.
- Determine the requirements for separating used grit from the residual cutting solution.

Most of the objectives listed above were met:

- Rockets were cut at appropriate locations.
- Explosives were successfully washed out, yielding particles that were small enough for subsequent hydrolysis. M28 rocket propellant could not be washed out, however, because of its tough, rubbery consistency. The propellant grain was separated as a single piece several times during the demonstration tests. In the full-scale operation, the technology provider proposes that the propellant grain would be separated, sheared, and the pieces shredded.
- Operational control during the cutting operations was demonstrated, although some refinements from original plans were necessary.
- Used grit was readily separated from the water used for cutting.

During demonstration tests, the propellant grain ignited and burned while it was being forcibly fed into the low-speed shredder. The committee had noted this possibility earlier (NRC, 1999):

> Friction, shear, or heat may result from the inadvertent introduction of metal, an excessive feed rate, or some other cause and could initiate the energetic material.

The problem was resolved during the demonstration by inundating the shredding face with cooling water. In the full-scale operation, the technology provider proposes performing the entire shredding operation under water.

The separation and initial processing of the rocket components (and other major hazardous operations) are performed remotely to reduce worker exposure to safe levels. Thus, although further development is necessary in certain

[1]WHEAT is an acronym for water hydrolysis of explosives and agent technology.

areas, the committee continues to believe that design and administrative controls will be feasible when the technology reaches the level of development at which quantitative risk assessments and hazard evaluations can be performed (NRC, 1999).

BIOTREATMENT SYSTEMS

Different biotreatment systems are used to treat HD hydrolysate and nerve agent (GB and VX) hydrolysates. Therefore, they are discussed separately below. The main objectives of the demonstration tests are listed below:

- to show a high level of destruction of Schedule 2 compounds and energetics hydrolysate compounds, yielding products acceptable for discharge
- to demonstrate a high level of destruction over an extended period of time with good operational control

The specific test objectives are listed below (DOD, 1999b):

- Validate that the immobilized cell biotreatment (ICB) process can eliminate Schedule 2 compounds present in all hydrolysate feeds.
- Confirm the absence of agent in the effluents of the ICB systems.
- Validate that the ICB systems (and the separately tested agent hydrolysis systems) can achieve a DRE of 99.9999 percent for VX, GB, and HD.
- Validate that the ICB systems (and the separately tested energetic hydrolysis systems) can achieve a DRE of 99.999 percent for energetics.
- Develop mass loading and kinetic data required for scale-up of ICB unit operations.
- Validate that the catalytic oxidation (CATOX) unit can eliminate specified VOCs, semi-VOCs, and Schedule 2 compounds from the process gas stream.
- Determine the potential impact of operating conditions on fouling and plugging of the CATOX unit.
- Characterize gas, liquid, and solid process streams from the ICB process for selected chemical constituents and physical parameters, as well as the presence or absence of agent, Schedule 2 compounds, and other toxic or hazardous compounds.

Biotreatment System for Mustard Hydrolysate

A flow diagram for the demonstration test unit used for HD hydrolysate is shown in Figure 4-1. The feed consisted

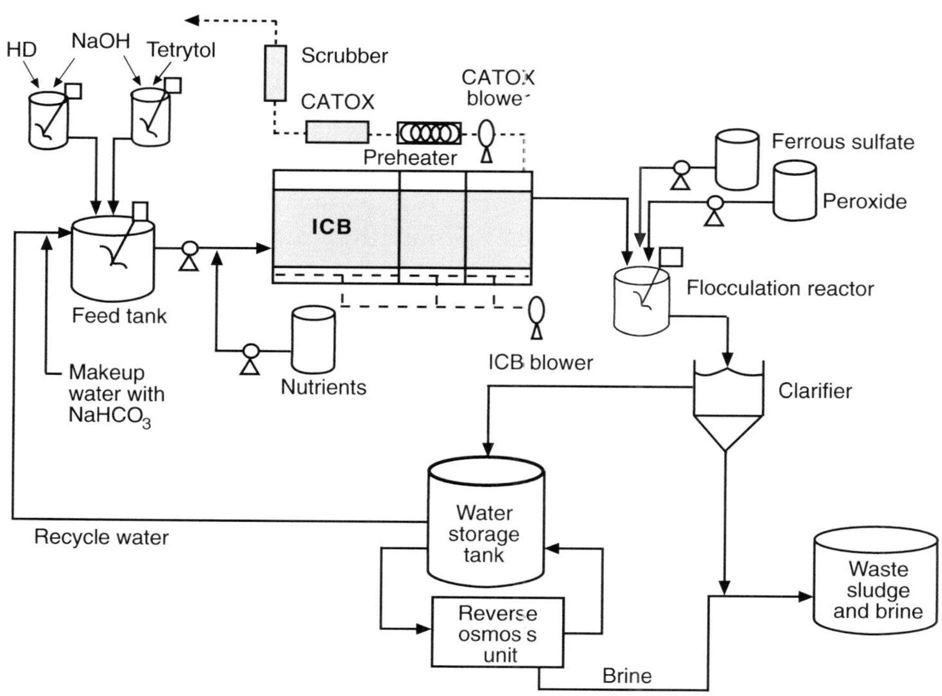

FIGURE 4-1 Demonstration test unit for treatment of HD/tetrytol hydrolysate. Source: Adapted from Parsons-AlliedSignal, 1999a.

of two streams, hydrolyzed HD and hydrolyzed tetrytol, that corresponded to the agent/energetics combination from an M60 105-mm artillery shell. Nutrients (ammonium sulfate or ammonium bicarbonate) were added to supply additional nitrogen. Fenton's reagent (30 percent H_2O_2 plus $FeSO_4$) was added to the flocculation product after biotreatment. (The amount added was small and did not contribute significantly to the overall oxidation process.)

The products of the HD biotreatment process are listed below:

- a wet sludge (biosolids plus a high percentage of brine)
- brine from the biotreatment system, concentrated by a reverse osmosis unit (which will not be included in the final plant design)
- depleted air from the bioreactors that has been treated in a CATOX unit (to oxidize organics carried over as spray or vapor) and then passed through a dry caustic scrubber and a carbon filter

The major criterion used to judge the efficacy of the bioreactor was the destruction of Schedule 2 compounds in HD hydrolysate and of tetrytol hydrolysate products in the feed. Schedule 2 compounds were not detected in the brine or sludge, and the DRE of Schedule 2 compounds from the combined agent/energetic hydrolysate feed was greater than 99.9 percent. Small amounts of 1,4-dithiane and 1,4-thioxane survived but were removed in the flocculation step, in which Fenton's reagent was added.

No energetics or their breakdown products, such as nitrotoluenes and nitrobenzenes, were detected in the brine and sludge, and samples of brine and sludge passed the EPA's toxic characteristic leaching procedure (TCLP). Low levels of several semi-VOCs were found in both sludge and brine samples. Also, some VOCs (e.g., 2-butanone and acetone) were found. Dioxins and furans were detected in some samples, but were below the levels of concern. Metals were also below the levels of concern. Analysis of the brine showed a greater than 90 percent removal of chemical oxygen demand (COD) relative to the bioreactor feed. Biological oxygen demand (BOD) in the brine was measured, but the technology provider considered the results meaningless because the microorganisms used in the BOD test were not well acclimated to the components in the agent and energetic hydrolysates.

Biotreatment System for Nerve Agent Hydrolysates

A flow diagram of the demonstration test unit used for nerve agent hydrolysate is shown in Figure 4-2. The main difference between this system and the system used for HD hydrolysate is the presence of the phosphonate form of phosphorus, which cannot be readily biodegraded. Other differences are listed below:

- A large amount of dextrose was added (about 44 lbs per pound of nerve agent products). The dextrose plus other nutrients (e.g., urea) represent about 95 percent of the total COD of the system.
- The feed rate of agent/energetics hydrolysate to the biotreatment system was scaled back to accommodate the added dextrose. The hydrolysate feed contained a concentration of about 0.1 percent Schedule 2 compounds, compared with 1.0 percent for the HD/energetics hydrolysate feed.
- The bio-oxidation process was augmented by a ultraviolet (UV)/hydrogen peroxide reactor.

The operating conditions were based on tests performed at the technology provider's laboratory that had demonstrated a DRE of more than 95 percent for Schedule 2 compounds. Energetics hydrolysate and agent hydrolysates were fed to the reactor in the following combinations:

- VX hydrolysate, Comp B hydrolysate, and M28 propellant (the products of a processed M55 rocket)
- GB hydrolysate and Comp B hydrolysate (the products of a processed M426 8-inch artillery shell)

The major criterion for judging the nerve agent process was the same as for the mustard process—destruction of the Schedule 2 compounds and energetic products in the hydrolysate feed. In general, the biotreatment of nerve agent hydrolysates was not successful. Although some of the problems were identified (described below), the reasons for the inadequate performance remain unclear.

The overall process (biotreatment plus UV/hydrogen peroxide) reduced the Schedule 2 compounds by 40 to 60 percent for GB hydrolysate, somewhat more for the VX hydrolysate (the higher DRE quoted in the demonstration report is an error [Parsons-AlliedSignal, 1999a].) As the test proceeded, there was a gradual buildup of the most biologically refractory Schedule 2 compounds: isopropyl methylphosphonic acid (IMPA) in GB hydrolysate; ethyl methylphosphonic acid (EMPA) in VX hydrolysate. The proportion of the DRE attributable to the biotreatment system and the proportion attributable to the UV/hydrogen peroxide could not be established from the available data. However, based on material balance estimates from the reported information on GB (for tests on April 13 and May 5), about 60 percent of the total IMPA conversion occurred in the bioreactor and flocculator, and about 40 percent occurred in the UV/hydrogen peroxide unit (Parsons-AlliedSignal, 1999a). The Fenton's reagent was not a major factor because the amount added corresponded to only about 3 percent of the initial feed COD.

The flow through the biotreatment process was continuous, 24 hours per day. However, the UV/hydrogen peroxide treatment was operated intermittently. A reverse osmosis

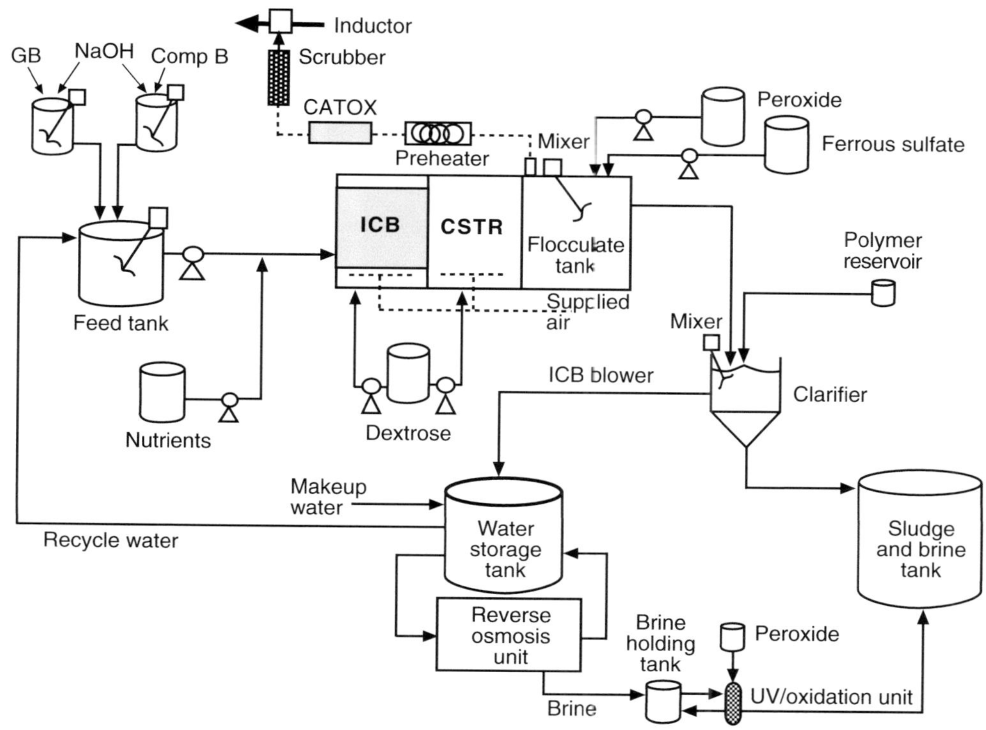

FIGURE 4-2 Demonstration test unit for treatment of GB/Comp B hydrolysate. Source: Adapted from Parsons-AlliedSignal, 1999a.

unit drawing on the recycle stream produced a small flow of concentrated brine, which accumulated in the brine holding tank. After 800 gallons had accumulated (over a period of about five days), the brine was recycled through the UV/hydrogen peroxide unit for six hours to produce one of the products leaving the plant. Although the UV/hydrogen peroxide unit was responsible for a substantial fraction of the total oxidation, a very large excess of hydrogen peroxide was used (at least 10-times the theoretical requirement). The intensity of the UV light was not reported. Therefore, the committee was unable to evaluate the efficiency of the UV/hydrogen peroxide unit.

The circulating brine was black, which suggests that the process was anaerobic in some areas. The black color-bodies were not identified, but, because of the black color, the brine was not suited for oxidation by the UV/hydrogen peroxide unit.

The air supply (116 cubic feet per minute [CFM]) was substantially less than planned (200 CFM) because of a much higher than expected pressure drop across the CATOX unit. The technology provider attributes this to an accumulation of corrosion products from the inlet line, but the CATOX unit was not examined (Lupton, 1999). The technology provider had obtained acceptable results with another unit when the air supply was equivalent to 200 CFM. Although a larger air supply might have resulted in acceptable levels of destruction during the demonstration tests, this is merely conjecture. Even at the reduced level of 116 CFM, the oxygen supply was more than 10 times the stoichiometric requirement (i.e., with air in at 21 percent oxygen, air out contained 19.5 percent oxygen). The reduced air flow might have caused poor dispersion of air in the reactor or uneven mixing and stirring. Nevertheless, the technology provider should examine the CATOX unit thoroughly and reassess its design.

During demonstration, the BOD of the hydrolysate feed was unusually large (15,800 mg/L)—in fact, 85-fold larger than the BOD during prior tests (200 mg/L). No explanation was given for the very large BOD requirement, which was much larger than the calculated oxygen requirement for complete oxidation of the feed. The very small BOD in the earlier tests was much lower than the theoretical oxygen requirement and undoubtedly much lower than the oxygen actually consumed in the tests. Too low a BOD measurement may be explained as a poor BOD test with a biological culture poorly acclimated to the feed, for example. However, there is no apparent explanation for a BOD measurement higher than the BOD for total oxidation. Therefore, the committee believes the BOD measurements are questionable.

"Thiol," the major Schedule 2 compound produced from VX hydrolysis, is expected to be more than 50 percent of the mass of Schedule 2 materials. Surprisingly, however, it is reported to be present in much smaller amounts, less than 5 percent in some cases. (To reduce the unpleasant odor, the hydrolysate may have been treated with sodium hypochlorite and the thiol oxidized, but no information on this is given. The treatment would have reduced the oxidation required of the bioreactor.)

According to the technology provider, the sludge produced in the biotreatment of nerve agents passed the TCLP tests satisfactorily. Because the treatment of these organophosphorus compounds was unsatisfactory, this sludge is not representative of the sludge that would be produced if the technology provider had developed an effective process. Thus, these TCLP tests are invalid.

The Army and its contractors have experienced problems with the analyses of the trace components in the effluents. For example, low concentrations of semi-VOCs and Schedule 2 compounds had to be measured in highly contaminated samples containing high concentrations of caustic. Many of the compounds in the effluents were never identified. Such solutions, as well as sludges, present difficult matrices in which to perform trace analyses. To lower the alkalinity, the samples were diluted extensively, thereby lowering further the concentrations of the trace components. Also, the caustic reacted with the absorbents, such as alumina, used in the chromatography columns (Arthur D. Little, Inc., 1999). In addition, the demonstration tests with the VX hydrolysate were delayed because the Army had difficulty analyzing the residual VX and certifying that the hydrolysate was safe prior to shipment.

The poor performance in the demonstration tests was attributed to the low air supply and the large BOD described above (Parsons-AlliedSignal, 1999b). However, the committee believes other factors may have been crucial:

- poor dispersion of air in the reactor leading to inadequate saturation of the liquid with oxygen
- inadequate acclimation of the biomass, particularly for handling phosphonate material

CATALYTIC OXIDATION

Catalytic Oxidation Unit for Mustard

The CATOX unit on the effluent gas from the biotreatment of HD hydrolysate appeared to work well, but because there were some difficulties in analyzing the gas for some EPA-regulated nonvolatile organic compounds, the gas composition was uncertain. The gas leaving the CATOX unit had traces of low molecular weight materials, which are considered acceptable. Chlorinated dioxins and furans were observed at very low levels in some of the analyses, but these compounds should be adsorbed from the gas by the carbon filter. No analysis of the gas discharged from the carbon filter was performed.

Catalytic Oxidation Unit for Nerve Agent

The technology provider claimed that the CATOX unit for the effluent gas from the biotreatment of GB and VX hydrolysates performed well. However, there was an unexpectedly large pressure drop across the unit (Lupton, 1999). Although both input and output streams were sampled, no data on the composition of the effluents were available.

METAL PARTS TREATER

The MPT system consisted of the following units:

- a cylindrical furnace heated electrically by induction heaters surrounding the cavity and by a flow of low-pressure superheated steam
- a furnace to generate and superheat the steam
- a heat exchanger to condense most of the outlet steam and the semi-VOCs
- a CATOX unit with a preheater and added air to treat gases leaving the condenser
- a solid caustic scrubber for the gas leaving the CATOX unit

The system was run in a batch mode. The chamber was loaded with the material to be treated, and the temperature and steam flow were ramped up to achieve a 5X decontamination condition. The chamber was cooled, opened, and the products examined. The proposed full-scale MPT system will have several batch reactors of the general type demonstrated, as well as a continuous reactor for some components (e.g., fuses and projectile burster casings). The continuous reactor will resemble the metal parts furnace of the baseline system but will be electrically and steam heated. The continuous processing unit was not included in the demonstration testing.

The MPT was tested with the following feeds:

- M60 105-mm projectile spiked with GB, VX, or HD
- dunnage of various kinds (wood, DPE suits, carbon)

To test a potential failure mode of the MPT, the CATOX unit associated with it was challenged by separately injecting 0.6 lb of each of the three neat agents over a four-hour period.

Qualitatively, the MPT system appeared to work well. Solid material remaining in the furnace was decontaminated to a 5X condition (free of agent). The condition of the material driven off in the furnace (the liquid and off-gas from the condenser) is more difficult to characterize. HD was reported to be less than 5 μg/L in the condensate; its breakdown (hydrolysis) products (thiodiglycol, 1,4-oxathiane, and dithiane) were at several hundred μg/L. In direct agent injection tests, GB was reported at 11 and 27 μg/L; VX at 60 and 220 μg/L. Their breakdown products were not reported. The volume flow of steam (and the volume of condensate) was not reported. Consequently, the breakdown level of the agents cannot be calculated.

A variety of VOCs were present in the condensate. The data on semi-VOCs were not available to the committee during the preparation of this report. There was some evidence of the reaction of organic materials with steam, but it was not possible to determine the extent or the weight fraction of feed material driven off in the furnace. The nature of the materials identified in the condensate suggests that they could be handled satisfactorily by recycling to the ICB reactor feed.

Neither analytical data on the gas from the condenser nor flow rates were included in the reports (DOD, 1999b; Parsons-AlliedSignal, 1999a). Some agent concentrations were reported based on analysis of depot area air monitoring system (DAAMS) tubes, which showed agent concentrations for VX from zero (i.e., not detected at the detection limit) to 25 times the time-weighted average (TWA) permissible exposure limit. (The committee assumed the TWA referred to was the stack-emission limit of 0.0003 mg/m^3).

During the direct injection tests, the CATOX unit destroyed the agent to a DRE of greater than 99.9999 percent. The very low levels of agent leaving the MPT unit should be destroyed in the CATOX (Parsons-AlliedSignal, 1999a).

The operational problems listed below must be addressed prior to the development of a final plant design:

- Some dunnage (e.g., DPE suits) generated gas too rapidly, resulting in an excessive temperature rise in the CATOX unit.
- Paint chips clogged the condenser liquid outlet.
- Some significant operational data were not reported: steam flow rate, liquid condensate rate, and vapor and air flow rates to the CATOX unit.
- The catalytic activity of the CATOX unit is expected to decline slowly with time, but this was not investigated during the demonstration tests.

It seems likely to the committee that the system performed its desired function. However, going from the batch, "unsteady-state" operation of the demonstration test unit to the proposed, "steady-state" flow operation will require further investigation by the technology provider.

SAFETY CONCERNS

The safety issues were discussed in the section on munitions-cutting and fluid-mining.

REEVALUATION OF STEPS REQUIRED FOR IMPLEMENTATION

The committee's earlier report identified seven steps required for implementation of the WHEAT technology (NRC, 1999).

The following steps would have to be taken to implement this technology package:

1. demonstration of the effectiveness of the biotreatment of various combinations of agent and energetics hydrolysates of sufficient length to give reasonable assurance of long-term performance

Mustard and energetics hydrolysates were effectively treated by the biotreatment process. However, nerve agent hydrolysates, which were mixed with energetic hydrolysates, were not digested by the microorganisms.

2. operation of the bioreactor at the planned salt-content

The demonstration tests were all done at very low salt contents (e.g., 0.5 weight percent). Other experience suggests that much higher salt contents could be tolerated (e.g., 2 weight percent [Lupton, 1999]).

3. characterization of the off-gas from the bioreactor to evaluate the extent of air-stripping from the reactor and the possible poisoning of the catalyst in the catalytic oxidation unit

This was not done during the demonstration tests (at least partly because of analytical difficulties). Therefore, the extent and rate of catalyst poisoning have yet to be determined. The extent of air stripping was not evaluated.

4. demonstration of the effectiveness and long-term performance of the catalytic oxidation system in destroying organic constituents in the bioreactor off-gas

Although the CATOX units appeared to perform well during the demonstration tests, their long-term performance remains to be demonstrated (see commentary on Step 3 above).

5. quantification and characterization of the sludge from the biological process to ascertain if Schedule 2 compounds or other hazardous constituents are present

The sludge from HD hydrolysates was tested and appeared to be nonhazardous. The sludge from the nerve agent hydrolysates also appeared to be acceptable, but they were the products of an operation that will require further development to perform satisfactorily and should, therefore, be retested as the system advances.

6. demonstration of unproven steps in the proposed process, including ultraviolet/peroxide oxidation and evaporation operations

Some "unproven steps" were demonstrated (e.g., high-pressure water-jet mining of explosives). Because the UV/hydrogen peroxide process was tested under very adverse conditions, its ultimate operation could not be evaluated. No brines were evaporated.

7. quantification and characterization of the salts from the evaporation operations to ascertain what organic compounds are present

Many partially oxygenated compounds were identified in the brine—some appeared in the dried salts and some evaporated in the drying operation. The compounds observed in the mustard/energetics process were materials typically observed in biotreatment plants. The nerve agent/energetics products, however, were the products of an unsatisfactory operation. Therefore, no conclusions can be drawn.

Because the demonstration test program was short, and because difficulties were encountered, few of the steps noted above were of sufficient duration to demonstrate long-term performance. However, it is reasonable to conclude that the biotreatment process will operate satisfactorily for HD hydrolysate. Because the nerve agent demonstration tests encountered many problems, further scale-up should be delayed until these problems have been resolved.

REVIEW OF PREVIOUS COMMITTEE FINDINGS

The demonstration program was responsive to some, but not all, of the committee's earlier findings.

> **Finding PA-1.** The biological treatment operation will require further demonstration to prove its ability (1) to handle a variety of feed stocks with reasonable acclimation times between changes, and (2) to achieve high levels of conversion of the Schedule 2 compounds in the hydrolysate. The demonstration will have to last long enough to give confidence in the long-term operational ability of the process.

The conversion rate of Schedule 2 compounds in the biotreatment process on mustard hydrolysate was high. Although acclimation time was longer than anticipated, this does not represent a serious problem. The results of the biotreatment of nerve agent hydrolysates, however, were discouraging. In both cases, the demonstration tests were too short in duration to demonstrate conclusively long-term operational reliability.

> **Finding PA-2.** The relative effects of biological treatment and air-stripping on the destruction of organic materials in the bioreactor have not been established. This will affect the composition of the off-gas from the bioreactor.

Air stripping was not seriously examined in the demonstration tests. However, the concentration of organics in the off-gas from the reactor was low.

> **Finding PA-3.** The effectiveness of ultraviolet/hydrogen peroxide oxidation in reducing Schedule 2 compounds to an acceptably low level has not been demonstrated. [Note: Applicable only to biotreatment of nerve agent hydrolysate.]

The UV/hydrogen peroxide process was operated under adverse conditions in the demonstration tests (i.e., the fluid was black and nontransparent).

> **Finding PA-4.** The bioreactor has been operated only at very low salt concentrations. Operation at design concentrations has not been demonstrated.

The demonstration reactions were also carried out only at low salt concentrations.

> **Finding PA-5.** Additional data should be gathered on the effectiveness of the catalytic oxidation system in destroying organic materials in the biotreatment of off-gas.

The CATOX system demonstrated high conversion of nerve agents and very low levels of materials in the off-gas. Input concentrations were very low, however, so the DRE could not be computed.

> **Finding PA-6.** The sludge from the biological process has not been completely characterized.

The sludges in the demonstration tests were extensively characterized. However, in contrast to the sludges produced from treatment of mustard hydrolysate, the sludges produced from biotreatment of nerve agent hydrolysates were not considered representative of a final acceptable process because of difficulties in processing the phosphonate form of phosphorus.

> **Finding PA-7.** Even though the evaporation operations involve conventional technologies, they have not been tested for this application.

No evaporation process was demonstrated.

> **Finding PA-8.** The dried salts from the evaporation operations have not been characterized for leachability and toxicity.

No dried salts were produced. Therefore, whether or not the dried salts will meet leachability and toxicity requirements for disposal, either with or without stabilization, was not determined.

SUPPLEMENTAL FINDINGS AND RECOMMENDATION

As a result of the demonstration tests, the committee's earlier findings (discussed above) have been supplemented by two new findings and a new recommendation:

Finding PA-1. The mustard demonstration tests were very encouraging and showed that the process is ready for the next scale-up.

Finding PA-2. The nerve agent demonstration tests had serious problems. However, if the previous tests at the technology provider's laboratory and the results of the demonstration tests are combined, the aggregate results are inconclusive. The reason for the poor demonstration results might be as simple as poor aeration in the bioreactor (see Recommendation PA-1).

Recommendation PA-1. Before proceeding to a further scale-up of GB and VX biotreatment processing, the committee recommends that the following steps be taken:

- The biotreatment process should be examined carefully at bench scale to determine the factors that are critical to success.
- An investigation of analytical techniques should be undertaken to provide more reliable process information.

5

Update of General Findings and Recommendations

Chapter 11 of the committee's initial report, *Review and Evaluation of Alternative Technologies for Demilitarization of Assembled Chemical Weapons*, included 16 general findings and seven general recommendations (NRC, 1999). For the most part, these findings and recommendations remain unaffected by the results of the demonstration tests of the three technology packages. Each of these findings and recommendations is quoted below followed by a discussion of the effect of the demonstration tests results. New findings are then presented.

REVIEW OF EARLIER FINDINGS AND RECOMMENDATIONS

> **General Finding 1.** The chemistries of all four of the primary technologies, (hydrolysis, SILVER II, plasma arc, and SET) as proposed, can decompose the chemical agents with destruction efficiencies of 99.9999 percent. However, each technology package raises other technical issues that must be resolved. One of the crucial issues is the identity and disposition of by-products.

Two of the three technology packages chosen for ACWA demonstration rely on hydrolysis as the primary treatment process. The third is based on plasma arc technology. Hydrolysis of agents was not a direct part of the demonstrations. However, the PMACWA produced approximately 1,100 gallons of GB hydrolysate, 400 gallons of VX hydrolysate, and 4,200 gallons of HD hydrolysate as starting materials for the demonstrations. The Army's ability to produce agent hydrolysates that show no agent above detection limits confirms the effectiveness of hydrolysis in destroying both mustard and nerve agents to a DRE of 99.9999 percent.

Although the Burns and Roe team did not conduct demonstration tests for the destruction of neat chemical agents, the committee continues to believe that a properly engineered plasma arc device could destroy both mustard and nerve agents to a DRE of 99.9999 percent.

> **General Finding 2.** The technology base for the hydrolysis of energetic materials is not as mature as it is for chemical agents. Chemical methods of destroying energetics have only been considered recently. Therefore, there has been relatively little experience with the alkaline decomposition of ACWA-specific energetic materials (compared to experience with chemical agents). The following significant issues should be resolved to reduce uncertainties about the effectiveness and safety of using hydrolysis operations for destroying energetic materials:
>
> - the particle size reduction of energetics that must be achieved for proper operation
> - the solubility of energetics in specific alkaline solutions
> - process design of the unit operation and the identification of processing parameters (such as the degree of agitation and reactor residence time) necessary for complete hydrolysis
> - the characterization of actual products and by-products of hydrolysis as a function of the extent of reaction
> - the selection of chemical sensors and process control strategies to ensure that the unit operation following hydrolysis can accept the products of hydrolysis
> - development of a preventative maintenance program that minimizes the possibility of incidents during the cleanup of accumulated precipitates

> **General Finding 3.** The conditions under which aromatic nitro compounds, such as trinitrotoluene (TNT) or picric acid, will emulsify in the aqueous phase and not be completely hydrolyzed are not well understood. Therefore, this type of material could be present in the output stream from an energetic hydrolysis step.

> **General Finding 4.** The products of hydrolysis of some energetic materials have not been characterized well enough to support simultaneous hydrolysis of different kinds of energetic materials in the same batch reactor.

> **General Recommendation 5.** Whatever unit operation immediately follows the hydrolysis of energetic materials should be designed to accept emulsified aromatic

nitro compounds, such as TNT or picric acid, as contaminants in the aqueous feed stream. (See General Finding 3.)

General Recommendation 6. Simultaneous processing of different types of energetic materials should not be performed until there is substantial evidence that the intermediates formed from the hydrolysis of aromatic nitro compounds will not combine with M28 propellant additives or ordnance fuze components to form extremely sensitive explosives, such as lead picrate. (See General Finding 4.)

The hydrolysis of energetics performed during the ACWA demonstrations substantiates the findings and recommendations cited above. The committee is concerned that the technology for the hydrolysis of energetic materials may be even more immature than was originally anticipated. Problems were experienced in scale-up test runs for Comp B and tetrytol that were not apparent during laboratory-scale tests. Because the hydrolysis of lead stearate produces lead hydroxide, toxicity is a potential problem, justifying the committee's concern about the possible formation of lead picrate if M28 propellant is simultaneously hydrolyzed with Comp B or tetrytol bursters.

The Army and General Atomics have acknowledged that more work needs to be done before the hydrolysis of energetics can be considered safe and effective at production-scale levels (DOD, 1999c, 1999d; General Atomics, 1999a). The demonstration tests provided a large body of data. The Army has assembled a team of agencies to analyze the preliminary results, assess the efficacy of the processes, and identify problems and their causes and effects. Further experimentation is also being planned.

General Finding 5. The primary chemical decomposition process in all of the technology packages produces environmentally unacceptable reaction products. Therefore, all of the packages are complicated processes that include subsequent treatment step(s) to modify these products.

The General Atomics and the Parsons-AlliedSignal technology packages use hydrolysis for primary chemical decomposition, whereas the primary treatment process in the Burns and Roe package is the PWC. Analyses of the agent hydrolysates produced by the PMACWA for the demonstration tests confirmed General Finding 5. In other words, the hydrolysates contained Schedule 2 compounds and other products that are not suitable for direct discharge to the environment. The PWC used for the demonstration was not tested on agents or under conditions that produced acceptable synfuel. Consequently, PWC by-products produced from agents must still be characterized.

General Finding 6. The waste streams of all of the ACWA technology packages could contain very small amounts of hazardous substances (besides any residual chemical agent). These substances were not fully characterized at the time of this report; therefore, all waste streams must be characterized to ensure that human health and the environment are protected. If more than one phase (gas, liquid, or solid) is present in a waste stream, each phase should be characterized separately.

Although a large body of data was gathered, the tests of unit operations from the three technology packages during the demonstrations were of short duration and were conducted with undersized reactors. In addition, the operating conditions were not optimized. Thus, the effluents that were produced may not be completely representative of the effluents that would be produced in units operating at different conditions (e.g., temperature, pressure, etc.).

General Finding 7. None of the proposed technology packages complies completely with the hold-test-release concept for all gaseous effluents (both process and ventilation effluents).

General Finding 8. Hold-test-release of gaseous effluents may not ensure against a release of agent or other hazardous material to the atmosphere. No evidence shows that hold-test-release provides a higher level of safety than current continuous monitoring methods for gaseous streams with low levels of contamination. Furthermore, none of the technologies provides for hold-test-release of effluents from ventilation systems that handle large volumes of gases from contaminated process areas.

Because the basic configurations of the three demonstrated technology packages have not changed, General Findings 7 and 8 remain unchanged. Hold-test-release was not included in the demonstration tests.

General Finding 9. Solid salts will be hazardous waste, either because they are derived from hazardous waste...or because they leach heavy metals above the levels allowed by the Resource Conservation and Recovery Act Toxicity Characteristic Leaching Procedure. Stabilization—mixing waste with a reagent or reagents to reduce the leachability of heavy metals—will probably be required before the salts can be sent to a landfill. The potentially high chloride and nitrate content of these salts will make the waste difficult to stabilize, and treatability studies will be necessary to determine a proper stabilization formula.

General Finding 9 remains unchanged because the demonstrations did not test the ability of unit operations to produce dried salts.

General Finding 10. Testing, verification, and integration beyond the 1999 demonstration phase will be necessary because the scale-up of a process can present many unexpected challenges, and the ACWA demonstrations were limited in nature.

The committee considers the demonstration tests as "proof-of-concept" tests of the demonstrated unit operations. In nearly all cases, the conditions during the tests had to be modified in some respects, and, in many cases, significant alterations had to be made to the procedures. Finally, the critical step of integrating the unit operations has not yet been addressed by the technology providers for any of the demonstrated technology packages.

The committee believes that the following general findings and recommendations from the committee's original report were not affected in any way by the demonstration tests of the three technology packages.

General Finding 11. Although a comprehensive quantitative risk assessment (QRA), health risk assessment (HRA), and ecological risk assessment (similar to assessments performed for the baseline process) cannot be completed at this stage of process development, these assessments will have to be performed and refined as process development continues.

General Finding 12. The "optimum" system for a particular chemical weapons storage depot might include a combination of unit operations from the technology packages considered in this report.

General Finding 13. Some of the ACWA technology providers propose that some effluent streams be used commercially. New or modified regulations may have to be developed to determine if these effluent streams can be recovered or reused.

General Finding 14. An extraordinary commitment of resources will be necessary to complete the destruction of the assembled chemical weapons stockpile in time to meet the current deadline using any of the ACWA technology packages. This would demand a concerted national effort. It is unlikely that any of the technology packages could meet this deadline.

General Finding 15. The Dialogue process for identifying an alternative technology is likely to reduce the level of public opposition to that technology. The committee believes that the Dialogue has been and continues to be a positive force for public acceptance of alternatives to incineration. Although the Dialogue process requires a significant commitment of time and resources, it has been a critical component of the ACWA program to date.

General Finding 16. Although the committee did not have access to scientific data on the attributes of a technology that would be most acceptable to the public, input from members of the active publics and previous research indicates that technologies with the following characteristics are likely to stimulate less public opposition:

- minimal emissions, particularly gaseous
- continuous monitoring of effluents to verify that the process is operating as designed (process assurance measurement)
- provisions for representatives of the local community to observe and participate in the process assurance measurement

General Recommendation 1. If a decision is made to move forward with any of the ACWA technology packages, substantial additional testing, verification, and integration should be performed prior to full-scale implementation (see General Finding 10).

General Recommendation 2. The sampling and analysis programs at each phase of development should be carefully reviewed to ensure that the characterization of trace components is as comprehensive as possible to avoid surprises in the implementation of the selected technology (see General Finding 6).

General Recommendation 3. If a decision is made to move forward with any of these technology packages, health and safety evaluations should progress from qualitative assessments to more quantitative assessments as the process design matures. Quantitative (QRA), health (HRA), and ecological risk assessments should be conducted as soon as is practical. Early initiation of these assessments will allow findings to be implemented with minimal cost and schedule impact (see General Finding 11).

General Recommendation 4. Any of these technology packages, or any component of these technology packages, should be selected on a site-specific basis (see General Finding 12).

General Recommendation 7. The Department of Defense should continue to support the Dialogue throughout the current ACWA program and should seriously consider the participation of the Dialogue in follow-on programs.

SUPPLEMENTAL GENERAL FINDINGS

In the Statement of Task for this report, the committee was asked to determine if any of the technology packages chosen for demonstrations was "viable to proceed with implementation of a pilot-scale program that would employ any of these technologies." The committee has evaluated the maturity of each unit operation in the proceeding chapters of this report. Table 5-1 provides a summary of the committee's assessments.

General Finding 1. Based on the committee's assessment of the maturity of the various unit operations (as summarized in Table 5-1), none of the three technology packages is ready for *integrated* pilot programming, although certain unit operations are sufficiently mature to bypass pilot testing (e.g., hydrolysis of agent).

The demonstrated PWC system of the Burns and Roe technology package does not appear to be ready for pilot testing for any assembled chemical weapons materials. The demonstrated components of the General Atomics technology package are close to achieving an overall acceptable level of maturity. However, certain key demonstration tests were not performed or the results were inconclusive. The demonstrated components of the Parsons-AlliedSignal technology package are also close

TABLE 5-1 Summary Evaluation of the Maturity of Demonstrated Unit Operations and Processes[a]

	Hydrolysates			Agent Munitions			
Unit Operation/Process	VX/GB	HD	Energetics	VX/GB	HD	Energetics	Other
Burns and Roe							
Plasma waste converter[b]	C	C	D	D	D	E	C[c, d,e]
General Atomics							
Hydrolysis				A	A		
Rotary hydrolyzer						C	
Shredding/hydropulping							A[c]
SCWO	B	B	C				C[c]
Parsons-AlliedSignal							
Munitions accessing				B	B	B	
Hydrolysis				A	A	C	
Biotreatment	D	A	A				
Catalytic oxidation							B[e]
Metal parts treater				B	B	D	B[d]

Note: Environmental and safety issues were considered in assigning maturity categorizations. Schedule and cost issues were not considered.

[a] The letter designations are defined as follows (a blank space indicates categorization was not applicable for that material).
- A Demonstration provides sufficient information to allow moving forward to full-scale design with reasonable probability of success.
- B Demonstration provides sufficient information to allow moving forward to the pilot stage with reasonable probability of success.
- C Demonstration indicates that unit operation or process requires additional refinement and additional demonstration before moving forward to pilot stage.
- D Not demonstrated; more R&D required.
- E Demonstrated unit operation or process is inappropriate for treatment.

[b] Includes integrated gas polishing system to support demonstration
[c] Dunnage
[d] Metal parts
[e] Effluents

to being ready for pilot testing, but only for mustard-bearing munitions. Hydrolysis of agent (used in the General Atomics and Parsons-AlliedSignal technology packages) appears sufficiently mature to consider full scale application to any assembled chemical weapons. Similarly, biotreatment of hydrolysate (Parsons-AlliedSignal technology package) appears sufficiently mature for full scale application to mustard munitions.

General Finding 2. The demonstration tests were not operated long enough to demonstrate reliability and long-term operation.

The PMACWA's demonstration plan was severely constrained by both scheduling deadlines and available budget resources. The technology providers did not have enough time for systemization (preoperational testing). Consequently, the committee maintains that these tests were simply "proof-of-concept" demonstrations that indicate whether or not a particular unit operation (with more development) might be applicable to the disposal of assembled chemical munitions.

General Finding 3. The committee reiterates that none of the unit operations has yet been integrated into a complete system. The lack of integration is a major concern and a significant obstacle to full-scale implementation.

References

Arthur D. Little, Inc. 1999. PMACWA Demonstration Test Data, Quality Review Summary, Parsons/AlliedSignal. 38652-20. Cambridge, Mass.: Arthur D. Little Inc.

Burns and Roe. 1999a. Assembled Chemical Weapons Assessment Program Final Report. June 1999. Oradell, N.J.: Burns and Roe Enterprises, Inc.

Burns and Roe. 1999b. Burns and Roe/Startech/Foster Miller Final Report Comments. Aberdeen Proving Ground, Md.: Program Manager for Assembled Chemical Weapons Assessment.

DOD (U.S. Department of Defense). 1997. Assembled Chemical Weapons Assessment Program Annual Report to Congress. December 1997. Aberdeen Proving Ground, Md.: Program Manager for Assembled Chemical Weapons Assessment.

DOD. 1998. Assembled Chemical Weapons Assessment Program Annual Report to Congress. December 1998. Aberdeen Proving Ground, Md.: Program Manager for Assembled Chemical Weapons Assessment.

DOD. 1999a. Assembled Chemical Weapons Assessment Program Supplemental Report to Congress. 30 September 1999. Aberdeen Proving Ground, Md.: Program Manager for Assembled Chemical Weapons Assessment.

DOD. 1999b. ACWA Final Technical Evaluation: Burns and Roe Plasma Arc; General Atomics Neutralization/SCWO; Parsons/Allied Signal Neutralization/Biotreatment. Appendix B *of Assembled Chemical Weapons Assessment Program Supplemental Report to Congress*. 30 September 1999. Aberdeen Proving Ground, Md.: Program Manager for Assembled Chemical Weapons Assessment.

DOD. 1999c. Comp B and Tetrytol Hydrolysis Milestone Report. Draft interim report on the hydrolysis process at the Pantex Plant. October 1999. Aberdeen Proving Ground, Md.: Program Manager for Assembled Chemical Weapons Assessment.

DOD. 1999d. M28 Hydrolysis Milestone Report. Draft interim report on the hydrolysis process at the Radford Army Ammunition Plant. October 1999. Aberdeen Proving Ground, Md.: Program Manager for Assembled Chemical Weapons Assessment.

EPA (Environmental Protection Agency). 1998a. Hazardous Waste Combustors; Revised Standards. Final Rule - Part 1: RCRA Comparable Fuel Exclusion. Permit Modifications for Hazardous Waste Combustion Units, Notification of Intent to Comply, Waste Minimization and Pollution Prevention Criteria for Compliance Extensions. June 19, 1998. Washington, D.C.: Environmental Protection Agency.

EPA. 1998b. Human Health Risk Assessment Protocol for Hazardous Waste Combustion Facilities. EPA530-D-98-001B. July 1998. Available on line at: http://www.epa.gov/epaoswer/hazwaste/combust.htm

General Atomics. 1999a. Assembled Chemical Weapons Assessment (ACWA) Draft Test Technical Report. June 30, 1999. San Diego, Calif.: General Atomics, Inc.

General Atomics. 1999b. Responses to PM ACWA Questions and Clarification Request: General Atomics Final Technical Report. Aberdeen Proving Ground, Md.: Program Manager for Assembled Chemical Weapons Assessment.

Lupton. 1999. Personal communication from Steven Lupton, PMACWA Office, to Walter May, Committee on Review and Evaluation of Alternative Technologies for Demilitarization of Assembled Chemical Weapons, October 22, 1999.

NRC (National Research Council). 1998. Using Supercritical Water Oxidation to Treat Hydrolysate from VX Neutralization. Committee on Review and Evaluation of the Army Chemical Stockpile Disposal Program. Washington, D.C.: National Academy Press.

NRC. 1999. Review and Evaluation of Alternative Technologies for Demilitarization of Assembled Chemical Weapons. Committee on Review and Evaluation of Alternative Technologies for Demilitarization of Assembled Chemical Weapons. Washington, D.C.: National Academy Press.

Parsons-AlliedSignal. 1999a. Assessment of Technologies for Assembled Chemical Weapons Demilitarization, Demonstration Test Final Report. July 1, 1999. Pasadena, Calif.: Parsons-AlliedSignal, Inc.

Parsons-AlliedSignal. 1999b. Responses to "Required" Comments: PM ACWA Questions and Clarification Request, July 14, 1999: Parsons-AlliedSignal Final Technical Report. Aberdeen Proving Ground, Md.: Program Manager for Assembled Chemical Weapons Assessment.

PMCD (Program Manager for Chemical Demilitarization). 1998. Fact Sheet: Demilitarization Protective Ensemble. Aberdeen Proving Ground, Aberdeen, Md.: Program Manager for Chemical Demilitarization.

Appendix A

Findings and Recommendations from the 1998 Report on Supercritical Water Oxidation

The following paragraph and the subsequent findings and recommendations are taken directly from *Using Supercritical Water Oxidation to Treat Hydrolysate from VX Neutralization* (NRC, 1998). They are reproduced here because the committee considers them applicable to the supercritical water oxidation (SCWO) technology evaluated in this study.

EXCERPT

Chemical neutralization of VX nerve agent results in the production of a liquid hydrolysate stream that has greatly reduced toxicity compared to the original nerve agent but requires further treatment to meet the requirements of the Chemical Weapons Convention and to be suitable for disposal. After considering several approaches, the U.S. Army has selected SCWO (supercritical water oxidation) as the primary process for treating the hydrolysate from VX neutralization prior to ultimate disposition. The integration of SCWO into the complete process for the destruction of VX stored at Newport, Indiana, also requires an evaporator system after SCWO treatment to allow water to be recycled back into the neutralization process. The evaporation system also produces a dry solid waste stream consisting of salts produced during the neutralization and SCWO treatment steps. Excess condensed water from the evaporator is expected to be of relatively high purity and suitable for discharge. The technology selected for the evaporation process step is mature with considerable full-scale design and operations experience. In contrast, treatment of the hydrolysate will be a new application for SCWO. Thus, the findings and recommendations presented here focus on the use of SCWO for the treatment of VX hydrolysate.

FINDINGS

Finding 1. Limited pilot-scale testing has demonstrated the ability of SCWO to achieve high destruction efficiencies for the organic constituents of VX hydrolysate. Effluent from SCWO treatment of VX hydrolysate has been shown to have negligible acute toxicity in intravenous testing in mice, gavage testing in rats, and dermal testing in rabbits. The separation of salts in the effluents from SCWO through an evaporator system should produce relatively pure water suitable for discharge and solid salts suitable for disposal. Treatment requirements for VX hydrolysate are less stringent than they are for VX because the hydrolysate has low toxicity relative to the agent. However, criteria for process destruction efficiency and final disposal standards have not been established.

Finding 2. Using SCWO to treat VX hydrolysate is significantly different and more complex than previous applications. SCWO systems on a pilot scale have been used to treat several other types of wastes, but SCWO is in commercial operation at only one site. There has been only limited pilot-scale or operational-scale experience with wastes that are similar to VX hydrolysate in being highly corrosive and salt-laden. Operation with VX hydrolysate or appropriate surrogates at design conditions, equipment configuration, or approximate scale for full-scale operations has not been demonstrated. A vertical cylindrical reactor is the only reactor configuration that has been successfully demonstrated to date at pilot scale for the treatment of VX hydrolysate and similar waste streams. Additional development and pilot-scale testing of SCWO technology will be necessary to ensure sustained, reliable operation of a full-scale integrated treatment system. Sufficient time appears to be available in the Army's implementation schedule for the Army to carry out development and testing for using SCWO at the Newport site, provided they are carried out expeditiously.

Finding 3. Pilot-scale operation of SCWO in a vertical cylindrical reactor at the temperature and pressure necessary for the effective destruction of hydrolysate constituents has been limited to one eight-hour and two less than two-hour tests. During pilot-scale testing with hydrolysate, the following factors were identified that could create difficulties in sustaining system performance:

- Large quantities of insoluble salts were produced, which must be effectively managed within, and downstream of, the SCWO reactor.

- Unexpected fluctuations were observed in temperature, pressure, and salt expulsion from the SCWO reactor.
- High levels of corrosion and erosion of materials of construction were observed in the reactor liner and pressure let-down valves.
- The sustained performance and reliability of the pressure let-down system was not demonstrated.

Although at this point in development the Stockpile Committee cannot be certain, it believes that a SCWO system for the treatment of VX hydrolysate with sufficient sustained performance can be achieved with additional development and testing.

Finding 4. Limited bench-scale and pilot-scale tests have demonstrated operating regimes under which SCWO can effectively destroy carbon-phosphorus bonds and oxidize the organic constituents present in VX hydrolysate. The demonstrated conditions for high levels of destruction (> 99 percent) include temperatures between 640°C (1184°F) and 730°C (1346°F) and pressures between 231 and 258 atm (3395 to 3792 psi). At temperatures and pressures below this regime, effluent from SCWO processing may contain significant concentrations of residual organic species that are difficult to destroy, including constituents with carbon-phosphorus bonds.

A basis for the reliable scale-up and operation of SCWO technology for the treatment of VX hydrolysate has not yet been demonstrated. Fundamental knowledge about the following processes within the SCWO reactor is still not available:

- the number and characteristics of the physical phases, including large quantities of entrained and adhered solids and potentially liquid, gas, and supercritical fluid phases
- fluid dynamics and mixing processes complicated by relatively high loadings of insoluble salts
- heterogeneous and homogeneous reaction mechanisms and kinetics
- salt nucleation, particle growth, agglomeration and adhesion mechanisms, and kinetics

Because the understanding of fundamental processes is limited and the process operational data and experience are sparse, empirical design and engineering judgment will be required for the selection of a prudent scale for development prior to full-scale demonstration. This is common engineering practice.

Finding 5. Alkaline VX hydrolysate and its destruction products under SCWO reaction conditions create an extremely corrosive and erosive environment that requires the careful selection of materials of construction. Although preliminary data indicate that certain noble metals, such as platinum and gold, may have acceptable properties, the data currently available are insufficient for the selection of materials of construction. The Army has initiated further testing of materials of construction.

Finding 6. Process monitoring and control strategies for the management of salts within the SCWO reactor and the destruction of the organic constituents of the hydrolysate have not been demonstrated.

RECOMMENDATIONS

Recommendation 1. A pilot-scale SCWO process facility with the critical characteristics of the full-scale design should be constructed and operated to further define operating characteristics and demonstrate sustained continuous operation of the process. Objectives for process development and demonstration should include:

- operation with either hydrolysate or a suitable surrogate to demonstrate reliable operation for periods similar to full-scale design operating cycles
- the development and validation of process monitoring and control strategies for salt management and the destruction of organic constituents
- the definition of stable operating regimes, including the temperature, pressure, and the use of the oxidant (liquid oxygen or compressed air) selected for full-scale operation
- the definition of a basis for process scale-up, operation, and maintenance of a full-scale system
- the development and demonstration of a reliable pressure let-down system

Because the understanding of the fundamental process mechanisms and operating characteristics is limited, the committee recommends that the pilot-scale system be within an order of magnitude of the total mass and heating throughput of a full-scale design unit. Based on testing and reactor scale-ups to date, a vertical cylindrical reactor configuration is recommended as the system that will probably require the least amount of additional development. Other reactor configurations may perform at required levels but would require significant additional development.

Recommendation 2. Testing of materials of construction should be carried out as necessary to finalize the selection of materials for critical components, including the SCWO reactor and the pressure let-down system. Additional pilot-scale testing indicated in Recommendation 1 should include fabrication with the materials of construction selected from testing smaller samples and evaluation of corrosion and erosion rates for critical components.

Recommendation 3. Flexibility and redundancy of critical components should be incorporated into the design of the full-scale system to allow for uncertainties about the basis for scale-up and operation. Trade-offs should be evaluated to establish an appropriate balance between two 100-percent capacity SCWO reactors or a greater number of smaller reactors. The analysis should consider performance uncertainties associated with process scale-up and complexity, as well as the reliability of operating several reactors in parallel.

Recommendation 4. The Army should make provisions for targeted research and development to resolve problems identified during pilot-scale testing and the full-scale implementation of SCWO technology.

Recommendation 5. Requirements for process destruction efficiencies and final disposal standards for all effluent streams from SCWO treatment should be clearly defined to ensure that the final design meets regulatory standards.

REFERENCE

NRC (National Research Council). 1998. Using Supercritical Water Oxidation to Treat Hydrolysate from VX Neutralization. Committee on Review and Evaluation of the Army Chemical Stockpile Disposal Program, Board on Army Science and Technology. Washington, D.C.: National Academy Press.

Appendix B

Biographical Sketches of the Committee Members

Robert A. Beaudet, *chair,* received his Ph.D. in physical chemistry from Harvard University. He joined the faculty of the University of Southern California in 1962 as an assistant professor and is now a full professor in the Chemistry Department. He has extensive knowledge of chemical-agent monitoring and detection technologies and has served on several Department of Defense committees on chemical-warfare agents.

Richard J. Ayen received his Ph.D. in chemical engineering from the University of Illinois. Dr. Ayen is currently an independent consultant and was formerly the director of technology for Waste Management, Inc. He has extensive experience in the evaluation and development of new technologies for the treatment of hazardous, radioactive, industrial, and municipal waste.

Joan B. Berkowitz graduated from the University of Illinois with a Ph.D. in physical chemistry. Dr. Berkowitz is currently managing director of Farkas Berkowitz and Company. She has extensive experience in the area of environmental and hazardous-waste management, a comprehensive knowledge of available technologies for the cleanup of contaminated soils and groundwater, and a strong background in physical and electrochemistry.

Nosa O. Egiebor graduated from Queens University in Kingston, Ontario, with a Ph.D. in mineral process and reaction engineering. Dr. Egiebor currently holds the Department of Energy Samuel P. Massie Chair of Excellence in Environmental Engineering at Tuskegee University. His areas of expertise span a broad range of topics in environmental engineering; his specific focus is on the biotreatment of hazardous wastes and supercritical-fluid technology.

Willard C. Gekler graduated from the Colorado School of Mines with a degree in petroleum-refining engineering. Mr. Gekler, formerly vice president, chief engineer at PLG, Inc., is currently an independent consultant. His extensive experience includes design and safety analysis of hazardous-materials handling, storage, and waste-treatment systems. He specializes in hazard evaluation, quantitative risk analysis, reliability assessment, and database development for risk and reliability.

Hank C. Jenkins-Smith received his Ph.D. in political science from the University of Rochester. Dr. Jenkins-Smith is currently a professor in the Department of Political Science at the University of New Mexico, where he is also the director of the Institute for Public Policy. His areas of expertise include statistical analysis, measurement of public opinion, politics of risk perception, environmental policy, and public policy.

John L. Margrave, a member of the National Academy of Sciences, graduated from the University of Kansas with a B.S. in engineering physics and a Ph.D. in physical chemistry. Dr. Margrave is currently the chief scientific officer at the Houston Advanced Research Center and the E.D. Butcher Professor of Chemistry at Rice University. His expertise is in high-temperature chemistry, materials science, and environmental chemistry.

Walter G. May, a member of the National Academy of Engineering (NAE), graduated with a Sc.D. in chemical engineering from the Massachusetts Institute of Technology. He was the senior science advisor for Exxon Research and Engineering Company from 1976 to 1983 and professor of chemical engineering at the University of Illinois from 1983 until his retirement in 1991. His expertise is in process design, thermodynamics, chemical-reactor design, separation processes, industrial chemistry and stoichiometry, and chemical-weapons disposal.

Kirk E. Newman received a B.S. in chemistry from the

College of William and Mary and an M.S. in chemical engineering from the University of Virginia. Mr. Newman is currently technology group leader for the Yorktown Detachment of the Naval Surface Warfare Center, Indian Head Division. He has extensive experience in the development, processing, and characterization of energetic materials used in military applications.

Jimmie C. Oxley received her Ph.D. in chemistry from the University of British Columbia in Vancouver and is currently an associate professor of chemistry at the University of Rhode Island. Her expertise is in thermal decomposition of energetic materials, explosives chemistry, and explosives safety.

William R. Rhyne received a B.S. in nuclear engineering from the University of Tennessee and an M.S. and D.Sc. in nuclear engineering from the University of Virginia. Dr. Rhyne is cofounder and director of H&R Technical Associates, Inc. He has extensive experience in risk and safety analysis associated with the processing and transport of hazardous nuclear materials and chemicals.

Stanley I. Sandler, a member of the NAE, graduated from the University of Minnesota with a Ph.D. in chemical engineering. Currently, he is the Henry Belin du Pont Professor and director of the Center for Molecular and Engineering Thermodynamics at the University of Delaware. His extensive research interests include applied thermodynamics and phase equilibrium, environmental engineering, and separations and purification.

William Randall Seeker, who received his Ph.D. in nuclear and chemical engineering from Kansas State University, is senior vice president of GE Energy and Environmental Research Corporation. Dr. Seeker has extensive experience in the use of treatment technologies and environmental-control systems for managing solid waste and controlling air pollution emissions. He is a member of the Executive Committee of the Environmental Protection Agency's Science Advisory Board.

Leo Weitzman received his Ph.D. in chemical engineering from Purdue University. He is a consultant with 28 years of experience in the development, design, permitting, and operation of equipment and facilities for treating hazardous wastes and remediation debris. Dr. Weitzman has extensive experience in the disposal of hazardous waste and contaminated materials by thermal treatment, chemical reaction, solvent extraction, biological treatment, and stabilization.